Petrographic Atlas: Characterisation of Aggregates Regarding Potential Reactivity to Alkalis

Isabel Fernandes · Maria dos Anjos Ribeiro
Maarten A.T.M. Broekmans · Ian Sims
Editors

Petrographic Atlas: Characterisation of Aggregates Regarding Potential Reactivity to Alkalis

RILEM TC 219-ACS Recommended Guidance AAR-1.2, for use with the RILEM AAR-1.1 Petrographic Examination Method

Springer

Editors
Isabel Fernandes
ICT, Institute of Earth Sciences and Department of Geology, Faculty of Sciences
University of Lisbon
Lisbon
Portugal

Maria dos Anjos Ribeiro
ICT, Institute of Earth Sciences and DGAOT, Faculty of Sciences
University of Porto
Porto
Portugal

Maarten A.T.M. Broekmans
Department of Mineral Resources
Geological Survey of Norway—NGU
Trondheim
Norway

Ian Sims
RSK Environment Ltd
Hemel Hempstead
UK

ISBN 978-94-017-7382-9 ISBN 978-94-017-7383-6 (eBook)
DOI 10.1007/978-94-017-7383-6

Library of Congress Control Number: 2015949285

Springer Dordrecht Heidelberg New York London

Printed on acid-free paper

Springer Science+Business Media B.V. Dordrecht is part of Springer Science+Business Media
(www.springer.com)

Foreword

This Petrographic Atlas is an essential adjunct to the RILEM Petrographic Examination Method AAR-1.1: Detection of Potential Alkali-Reactivity, which is itself an essential first step in undertaking a comprehensive assessment of an aggregate.

Over a number of years successive RILEM Technical Committees have developed recommendations to enable engineers, specifiers and testing houses to design and produce concrete which will not suffer damage arising from alkali reactions. The full suite of these recommendations has now been published by RILEM in Volume 17 of their 'State-of-the-Art Reports' series. In this report (Nixon and Sims 2016), there are five recommended methods of test for aggregates (designated AAR-1 to AAR-5) and an overall recommendation which describes how these should be used to enable a comprehensive aggregate assessment (AAR-0). Additionally, there are three Recommended International Specifications for concrete (AAR-7.1 to AAR-7.3), which describe how the aggregate assessment can be combined with other measures in the design of the concrete to produce concrete with a minimal risk of developing damage from alkali reactions.

When assessing an aggregate for use in concrete AAR-0 recommends that the first step should be to undertake a petrographic examination of the aggregate using AAR-1.1. This will enable the lithology of the aggregate to be established and aggregate to be classified as:

Class I—Very unlikely to be alkali-reactive
Class II—Alkali-reactivity uncertain
Class III—Very likely to be alkali-reactive

On the basis of this petrographic examination and classification decisions can be made, using the methodology shown in AAR-0, about the need for further testing and which test methods are appropriate. However, it is recognised that it can be difficult for petrographers working with rock types with which they are unfamiliar to identify the facets which may make that rock susceptible to alkali reactions. This Atlas, therefore, presents micrographs and descriptions of rocks from around the world which have been found to be involved in damaging alkali reactions. It is hoped that it will prove to be a valuable assistance in improving the accuracy of aggregate assessment.

The examples of rocks given in the Atlas are chosen as far as possible from structures affected by damaging ASR. The accurate identification of damage arising from ASR and the precise identification of the rocks, mineral constituents and textural features involved in this damage is an essential way of building confidence in petrography. Guidance on the correct diagnosis of AAR damage is given in the published State-of-the-Art Report for AAR-6.1 (Godart et al. 2013).

The successful completion of this Atlas depended on the assistance of many members of the RILEM TC, by providing samples and information, but above all it has only been possible because of the dedicated hard work of Isabel Fernandes and her team at the Universities of Porto and more recently Lisbon.

February 2016

Philip Nixon
Chairman RILEM TC 219-ACS

Acknowledgments

The editors are grateful to the following colleagues from different countries for their contributions in providing samples and/or reviewing of the texts:

Ahmad Shayan–Australia and New Zealand
António Santos Silva–Portugal and Angola
Benoit Fournier–Canada and USA
Bertie Oberholster–South Africa
Børge Wigum–Iceland
Bruno Godart–France
Catherine Wong Phui Chan–Singapore
Christine Merz–Switzerland
Colin Giebson–Germany
Duyou Lu–China
Eric Brouard–France
Esperanza Menéndez-Méndez–Spain
Geoffrey Blight (Deceased)–South Africa
Ingmar Borchers–Germany
Jan Lindgård–Norway
Josée Duschesne–Canada
Karin Appelquist–Sweden
Katrin Seyfarth–Germany
Mario de Rooij–The Netherlands
Mark Alexander–South Africa
Miguel Ferreira–Finland
Nélia Castro–PARTNER Project thin sections
Nicole Pagan Hasparyk–Brazil
Özge Andiç-Çakır–Turkey
Paul Bennett-Hughes–UK
Sigrun Kjær-Bremseth–Norway
Silvina Marfil–Argentina
Sylvine Guédon–France
Sue Freitag–New Zealand
Tetsuya Katayama–Japan and Iceland

This work has been developed in the scope of the activities of ICT, Institute of Earth Sciences, under the Strategic Project Pest-OE/CTE/UI0039/2014 and also the Projecto IMPROVE-PTDC/ECM/115486/2009, Improvement of performance of aggregates in the inhibition of alkali-aggregate reactions in concrete, funded by FCT (Fundação para a Ciência e

Tecnologia). The editors are grateful to Rita Lamas and Sara Leal for their contribution with the study of some of the samples, for organizing the thin sections and insertion of the scale bars.

The SEM/EDS results were obtained using the equipment available at CEMUP which were financed by REEQ/1062/CTM/2005 and REDE/1512/RME/2005 of FCT.

Contents

Abbreviations

AO	Angola
AR	Argentina
AU	Australia
BE	Belgium
BR	Brazil
CA	Canada
CH	Switzerland
CN	China
DK	Denmark
ES	Spain
FI	Finland
FR	France
GE	Germany
IS	Iceland
IT	Italy
JP	Japan
NA	Republic of Namibia
NL	The Netherlands
NO	Norway
NZ	New Zealand
PT	Portugal
SE	Sweden
SG	Singapore
TR	Turkey
UK	United Kingdom
USA	United States of America
ZA	Republic of South Africa

BSE	Back Scattered Electrons
EDS	Energy Dispersive Spectrometry
GPXPL	Crossed-Polarized Light with Gypsum Plate
PPL	Plane-Polarized Light
SEM	Scanning Electron Microscope/Microscopy
XPL	Crossed-Polarized Light

1 Introduction

Isabel Fernandes, Maarten A.T.M. Broekmans, Maria dos Anjos Ribeiro, and Ian Sims

Since deleterious alkali-silica reaction (ASR) in concrete was recognised for the first time in the 1940s, research has been developed in order to identify and understand the phenomena, materials, and mechanisms involved. ASR is a complex chemical reaction still not fully understood, involving alkalis dissolved in the concrete pore fluid and some forms of reactive silica when enough humidity is available. Deleterious ASR involves expansion within hardened concrete and thereby induces extensive cracking, providing easy access of the environment to the concrete interior, thus enabling physical deterioration mechanisms to accelerate overall structural decay.

RILEM AAR-0 2016 (Nixon and Sims 2016) recommends that the assessment of aggregates for use in concrete must start with petrographic examination, conforming with AAR-1.1 (developed by RILEM 219-ACS) to identify rock types present and their relative proportions. The modal content of any potentially reactive constituents is then used to classify the aggregate as:

Class I—Very unlikely to be alkali-reactive
Class II—Alkali-reactivity uncertain
Class III—Very likely to be alkali-reactive

Several countries have developed national standards for the petrographic assessment of concrete aggregates (e.g. BS 812-104 1994; BS 7943 1999; FD P 18-542:2004, NCA 2004, CUR Recommendation 89 2008, CSA A23.2-15A 2009; ASTM C294 2011, ASTM C295 2012) based mainly on the petrographic analysis of the aggregates under the optical microscope. Complementary methods are recommended when needed, in particular for very fine grained rocks.

The distinction between innocuous and potentially reactive aggregates is not strict but gradual, depending on the types of silica present (e.g. quartz, opal) and its geological history, causing various textures. Many years of practice with petrographic assessment of aggregates have made evident that maintaining a list of alkali-reactive rock types provides insufficient warranty against deleterious ASR. In addition, correct identification and classification of potentially alkali-reactive rock types still strongly relies on the petrographer's skill and experience with local geology.

Table 1.1 shows that the numbers of rock types described in widely accepted national standards varies considerably and goes into different detail. For example, the large number of sedimentary rocks considered by Canadian Standard CSA A23.2-15A is due to the division into altered varieties of igneous, metamorphic and sedimentary rocks.

This atlas takes a starting point in the European aggregate atlas prepared for the PARTNER project (Lorenzi et al. 2006), initiated by RILEM TC-191 ARP members. For the preparation of this atlas it was necessary to define what sample data should be included ideally, and finding a balance with what was practically feasible given available means and time. The atlas is intended to assist in the identification of potentially alkali-reactive rock types in aggregates for concrete. It presents micrographs of known and confirmed deleterious rock types to enable experienced petrographers to compare with their own thin sections under study. This atlas represents RILEM AAR-1.2, and is complementary to the petrographic examination method described in RILEM AAR-1.1.

I. Fernandes (✉)
ICT, Institute of Earth Sciences and Department of Geology, Faculty of Sciences, University of Lisbon, Lisbon, Portugal
e-mail: mifernandes@fc.ul.pt

M.A.T.M. Broekmans
Department of Mineral Resources, Geological Survey of Norway - NGU, Trondheim, Norway
e-mail: maarten.broekmans@ngu.no

M.A. Ribeiro
ICT, Institute of Earth Sciences and DGAOT, Faculty of Sciences, University of Porto, Porto, Portugal
e-mail: maribeir@fc.up.pt

I. Sims
RSK Environment Ltd, Hemel Hempstead, UK
e-mail: ian@simsdoc.com; isims@rsk.co.uk

I. Fernandes et al. (eds.), *Petrographic Atlas: Characterisation of Aggregates Regarding Potential Reactivity to Alkalis*, DOI 10.1007/978-94-017-7383-6_1

Table 1.1 Numbers of rock types listed in selected standards

standard/guideline	igneous	metamorphic	sedimentary	sum
EN 932-3	12	7	11	30
CSA A23.2-15A	17	12	>70	>99
ASTM C294, ASTM C295	11	9	9	29
BS 812-104/BS 7943	11	8	12	31
AAR-1.2 (this atlas)	8	12	5	25

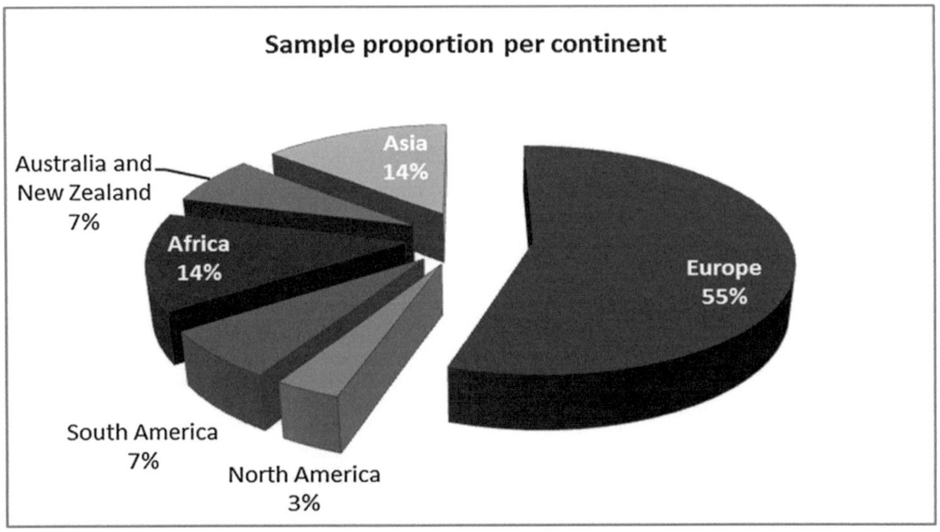

Fig. 1.1 Relative proportions per continent, in percent of samples. *Note* some countries had more than one atlas contributor

1.1 Sample Materials

Rock types identified as deleterious in concrete field structures have been collated in lists and tables (e.g. Wigum 2000; Oberholster 2009; RILEM AAR-1 2003), some short and concise, others long and extensive. These lists and tables reflect local (national, regional) experiences, based on locally available aggregate materials. However, the same rock types locally identified as alkali-reactive, have been found to be innocuous elsewhere, which may reflect differences in detailed petrographic properties and qualities, e.g. local geological history and petrogenesis, mineral modal content, structure, texture/fabric, etc., in addition to specific exposure conditions and construction practice.

Consequently, no single rock type uniformly classifies as alkali-reactive, or as innocuous. Therefore, whenever possible, classification of rock types in terms of alkali-reactivity demands application of local acceptance criteria, based upon experience and practice with the material under consideration.

This atlas was prepared from the petrographic thin section assessment of samples from a number of countries around the world, kindly supplied by RILEM TC-219 ACS members or their direct contacts in Argentina, Australia, Brazil, Canada, China, Finland, France, Germany, Iceland, Japan, Netherlands, New Zealand, Norway, Portugal, Singapore, South Africa, Spain, Sweden, Turkey, UK and USA. In addition, seven thin sections prepared from concrete prisms after laboratory expansion testing for the EU PARTNER Project were included, containing aggregates from Belgium, Denmark, Germany, Italy, Norway and UK. Figure 1.1 presents the relative proportion of samples per continent, as a percentage of all samples contributed. Sample provenance is heavily biased towards Europe, coinciding with RILEM-member provenance, and therefore might not be completely representative of 'world ASR rocks'.

Figure 1.2 presents a world map listing relative proportions of the main rock types analysed (i.e. igneous, sedimentary, metamorphic) per continent. In this calculation of the rock types classified as potentially reactive out of the samples analysed, it is clear that the statistical base improves with higher numbers of samples, as is the case for Europe.

Exclusively based on the sample materials assessed for this atlas, it can be concluded that the forms of silica behaving deleteriously in one given alkali-reactive rock type are rather consistent. Reactive forms of silica will be different in another rock type, but a certain overlap occurs. However, reliable identification of alkali-reactive and

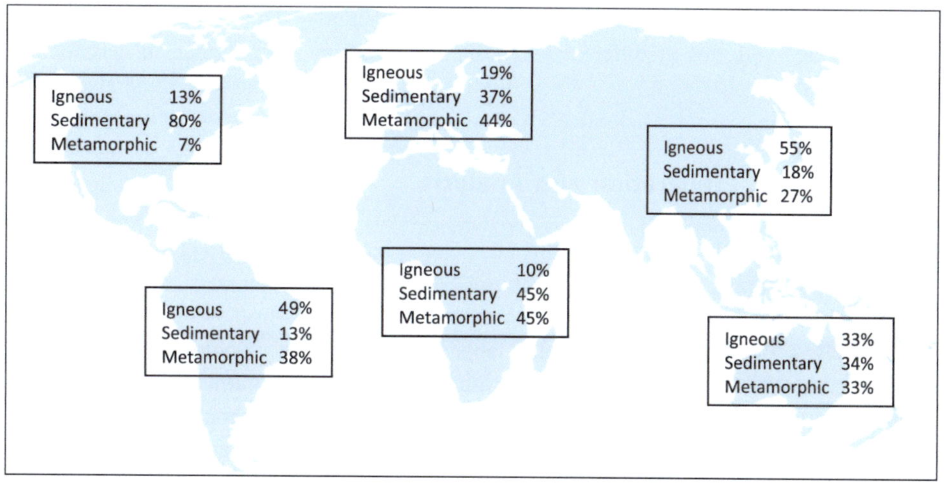

Fig. 1.2 Relative proportions per continent, of rock types represented in this atlas

innocuous varieties of a single rock type through optical petrographic assessment alone remains a challenge.

1.2 Disclaimer

This Atlas does not represent *all* potentially deleterious alkali-reactive rocks in the world, but only and exclusively rock types previously identified as alkali-reactive in concrete field structures or laboratory expansion testing, for which samples have been supplied.

Rock types not included in this atlas may still be potentially reactive. New aggregate resources are being identified and sampled every day for evaluation and assessment by petrography and additional laboratory methods, prior to prospective use in concrete. In addition, rock types not previously identified as alkali-reactive, including some slow-reacting forms, may be recognised as such in field structures, owing to aging of existing structures and to enhanced diagnosis of deleterious ASR.

Rock types classified as potentially reactive in this atlas may still be suitable for use in concrete, provided appropriate mitigation strategies are applied, as published in specialised literature (e.g. RILEM AAR-7.1, see Nixon and Sims 2016).

The authors do not accept any liability or legal responsibility for any errors or omissions that may be made. The authors make no warranty, expressed or implied, with respect to the material contained in this atlas.

1.3 Materials and Methods

1.3.1 General Considerations

This petrographic atlas is designed to assist in the correct identification and classification of rock types, by presenting thin section micrographs from selected rock types with confirmed and/or documented deleterious alkali-reactivity, namely from field structures, as these represent the most important confirmation of reactivity and from dependable laboratory expansion tests on concrete prisms or mortar bars. As an alternative, rock fragments of aggregate from well researched and established alkali-reactive resources were included. Colleagues listed in the Acknowledgements section also provided a number of thin sections of rocks and concrete. The atlas authors have assessed all sample materials supplied, using thin section petrography under a high quality optical microscope, occasionally supported by a scanning electron microscope (SEM).

Examples from concrete field structures included bridges, viaducts, dams, highway pavements, hydro-electric power stations, power line foundations, railway sleepers, retaining walls and airport aprons. A number of these have been described in literature, e.g. Batic et al. (1994), Oberholster (1996, 2009), Silveira et al. (2000), Hasparyk et al. (2004, 2008), Andrade et al. (2008), Leemann and Merz (2012a, b), Rogers and MacDonald (2012), Sanchez et al. (2012).

The assessment of alkali-reactive rock types in polymictic aggregate materials (unconsolidated sand and gravel of mixed composition) was carried out by preparing thin sections from selected areas in the concrete prism samples of the same materials showing features considered diagnostic for deleterious ASR, e.g. cracks parting aggregate particles and extending into the cement paste, ASR-gel filling cracks or air voids (French 1991; St John et al. 1998; also see Grattan-Bellew and Mitchell 2006).

Petrographic assessment of concrete aggregate is practically limited to the application of rock nomenclature based on geological classification criteria (examples in Table 1.1). However, rock nomenclature does not by any means reflect its engineering properties such as the behaviour in concrete, and prediction of performance properties based solely upon nomenclature must be avoided. For example, 'granite' is rightly regarded as non-reactive in some regions, whereas other 'granite' varieties in different areas are found to be notably reactive.

1.3.2 Specimen Preparation and Analysis

Thin sections specially produced for this atlas were prepared following procedures detailed in Humphries (1992) with minor adaptations and mainly without fluorescent dye. Thus, sized billets of 10 mm thick were cut and lapped plane, then mounted on $\sim$27 × 48 mm sized carrier glass with clear epoxy resin to produce 25 × 43 × 0.030 mm thin sections. After full setting, excess material was cut off with a diamond blade, the section surface impregnated with clear epoxy resin to reinforce the sample material, seal off porosity and reduce preparation artefacts. Sections were then manually lapped to desired thickness using carborundum powder in successively finer grading. Thickness was verified by petrographic microscope. Some thin sections of very-fine grained rocks were prepared thinner than the conventional 30 μm to minimise overlapping mineral grains which complicates identification.

Finished thin sections were by default covered with a glass slip or spray, unless SEM assessment was necessary, in which case the thin sections were polished. As above referred, instead of hand specimens, some colleagues sent ready-made thin sections. Therefore, some differences will be found in the images as some of those thin sections were already impregnated with yellow-fluorescent dye. Also, some variations can be found regarding the thickness of the thin sections.

Thin sections were assessed using a Nikon Eclipse E400POL© petrographic microscope in plane-polarized light (PPL) and crossed-polarized light (XPL). The gypsum plate (GPXPL) mode was used when the rock was very fine-grained in order to identify the mineral/vitreous composition, such as in cherts, for example. Photomicrographs were obtained using a Zeiss MRc5© camera and Axiovision© software. The presence of very-fine grained silica in siliceous limestones was confirmed using high resolution (Schottky) Environmental Scanning Electron Microscope with X-ray Microanalysis and Electron Backscattered Diffraction analysis Quanta 400 FEG ESEM/EDAX Genesis X4 M. Samples were coated with a carbon thin film, by vapour deposition, using JEOL JEE-4X Vacuum Evaporator equipment.

1.3.3 Thin Section Petrography

Rocks are classified using criteria based on mineral content, (micro-) structure, texture/fabric, and petrogenetic context. The modal content in rock-forming minerals is of key importance in classification, whereas presence of minor or accessory minerals facilitates distinction from similar rock types. Structure describes the overall spatial arrangement of the rock, e.g. layered, migmatized, vesicular, brecciated. Texture (also called fabric) describes in geometrical terms how the rock's constituent minerals are mutually intergrown, considering degree of crystallisation, grain size, shape, contacts and additional features (MacKenzie et al. 1982; Brodie et al. 2002; Vernon 2011).

Geological field information is essential for unequivocal identification of rock type and the correct application of nomenclature. However, field information may be lacking for samples from remote or foreign locations, or from resources rarely used or with uncertain geological setting; sometimes such information remains partially undisclosed for existing concrete with incomplete documentation. Considering the above, in addition to the application of rock nomenclature and the identification of potentially alkali-reactive forms of silica (i.e. quartz, chalcedony, opal, moganite, glass), this atlas pays special attention to various aspects of texture/fabric, including (but not limited to) serrate or interlobate grain boundaries, (sub-) grain size, undulatory extinction and deformation lamellae in quartz. These properties and qualities are thought to contribute to variable alkali-reactivity behaviour in concrete between different rock types, as well as for a single given rock type from one locality to another, laterally within a single geological formation, or from different levels within a single quarry (e.g. Oberholster 2009; Hünger 2005; Broekmans et al. 2009).

The captions with each atlas entry describe the potentially reactive forms of silica and/or textural features identified in that particular thin section. Quartz is identified as microcrystalline if <100 μm, as cryptocrystalline if <10 μm, after recommendations by Grattan-Bellew (1992) (also see Alaejos and Lanza 2011; Blight and Alexander 2011). In practice, this essentially coincides with the definition of MacKenzie et al. (1982): 'mineral grains that can be identified in thin section with the optical petrographic microscope are classified as microcrystalline, those too small are cryptocrystalline'.

1.4 Organization of the Atlas

The most reliable way to confirm alkali-reactivity of aggregate constituents is through thin section petrography of concrete from a damaged field structure, wherein sites of expansive reaction have demonstrated that damage relates to the reaction, and, alternatively, on prisms after completed and dependable laboratory expansion testing. However, in some cases the samples contributed consisted of fragments

Table 1.2 Overview of rock types included in this atlas

igneous	sedimentary	metamorphic
andesite	chert/flint	amphibolite
basalt	siliceous limestone	calc-silicate rock
dacite		granofels
felsic porphyry	greywacke	quartzite
granite	mudstone	
rhyodacite	sandstone	slate, phyllite, schist, gneiss
rhyolite		
tuff		*tectonites*: fault breccia, cataclasite, mylonite, pseudotachylite

of aggregate materials for which the reported reactivity potential could not be independently corroborated by petrographic thin section assessment, so that the sample was not considered for the atlas unless further information was provided by suppliers. The photomicrographs focus primarily on rock (micro-) structure and texture/fabric (less on mineral content); most images were taken in crossed polarised light (XPL) to reveal grain size and morphology of potentially alkali-reactive silica varieties which are mostly transparent and colourless under plane polarised light (PPL).

Table 1.2 collates the rock types presented in the atlas, with traditional division into igneous, sedimentary, and metamorphic lithologies. Rock classification and nomenclature follow the current recommendations from International Union of Geological Sciences—IUGS, namely Hallsworth and Knox (1999) for sedimentary rocks, Le Maitre et al. (2005) for igneous rocks and Fettes and Desmons (2007) for metamorphic rocks. Nevertheless, the authors are well aware that some names may deviate substantially from local names, which often refer to field information not visible in thin section or hand specimen, and not available to these authors. The situation in fact represents a 'blind assessment' where the origin of the material is undisclosed, which would be the most objective way to assess materials, instead of by purely instrumental methods.

Metamorphic rocks may have inherited structure and texture/fabric from their igneous, sedimentary, or metamorphic protolith, whereas mineral content and assemblage reflect their current metamorphic nature. In this Atlas, in order to keep it simple, the samples with slight features of metamorphic conditions for which the protolith could be determined were kept in the respective original rock type. The metamorphic features (e.g. texture) were just included in the sample description.

Table 1.3 Reactive silica forms commonly observed in rock types in this atlas

crystalline quartz			silica polymorphs[a]	volcanic glass incl. devitrified
microcrystalline <100 μm	cryptocrystalline <10 μm	undulatory or subgrained		
granite	rhyolite	granite	andesite	andesite
	rhyodacite		basalt	basalt
flint/chert		gneiss	dacite	felsic porphyry
greywacke	flint/chert	quartzite	rhyodacite	rhyolite
sandstone	greywacke	schist	rhyolite	
	mudstone			
granofels	sandstone	fault breccia	flint/chert	
phyllite	siliceous limestone	cataclasite	sandstone	
schist		mylonite	siliceous limestone	
slate	calc-silicate	pseudotachylite		
	gneiss			
	phyllite			
	quartzite			
	schist			
	slate			
	cataclasite			
	fault breccia			
	mylonite			
	pseudotachylite			
	tuff			

[a]*High temperature* cristobalite, tridymite; opal, chalcedony, moganite, low-quartz (α-SiO_2)
In each column, the igneous, sedimentary and metamorphic rock types are separately listed

Table 1.3 groups the same rock types according to the type of reactive silica present. It appears that the types of reactive silica occurring in a given rock type are rather consistent regardless of provenance. Also, cryptocrystalline quartz (<10 μm) is the most ubiquitous type of reactive silica, present in most rock types included in this atlas. By contrast, volcanic glass is only present in a much more limited number of rock types, as are the silica polymorphs. While awareness of this observation may assist in the diagnosis of potentially alkali-reactive rock types, it does not provide a basis for an alternative manner of re-grouping lithologies, as none of the reactive silica types is exclusive to a certain rock type.

After the petrographic descriptions of rocks and aggregates, selected images of ASR-damaged concrete, illustrating the features characteristic of reactivity in hardened concrete are presented. All literature consulted in the preparation of this atlas is collated in the reference list, including more elaborate descriptions of aggregate materials whose deleterious constituents are included in this atlas.

1.5 Summary

This RILEM AAR-1.2 Atlas is complementary to the petrographic method described in RILEM AAR-1.1. It is designed and intended to assist in the identification of alkali-reactive rock types in concrete aggregate by thin section petrography. Additional issues include:

- optical thin section petrography conforming to RILEM AAR-1.1 is considered the prime assessment method for aggregate materials, being effective regarding cost and time. Unequivocal identification of minerals in very-fine grained rock types may however require use of supplementary methods.
- the atlas adheres to internationally adopted schemes for rock classification and nomenclature, as recommended in AAR-1.1. Thus, rock types are classified as igneous, sedimentary or metamorphic based upon mineral content, microstructure, and texture/fabric.
- in addition, the atlas identifies known alkali-reactive silica forms in each rock type presented. It also identifies consistent coincidence between certain lithologies and silica types; however, it refrains from attributing alkali-reactivity to a specific silica property or quality.
- operator skill and experience remain essential for reliable assessment by thin section petrography.
- aggregate materials must be classified according to local criteria, based on regional experiences with ASR-damaged field structures and geology. Access to additional data may be relevant for the assessment of imported materials.
- mere application of rock nomenclature does not provide any sort of warranty to the development of deleterious alkali-reaction. Such may result in either rejection of a suitable aggregate material thus wasting a valuable resource, or acceptance of an unsuitable material leading to concrete damage, both of which are undesirable.

References

AFNOR FD P 18-542 (2004): Granulats: Critères de qualification des granulats naturels pour béton hydraulique vis-à-vis de l'alcali-réaction. Association Française de Normalisation, Paris, France.

Alaejos, P, Lanza, V (2011): Influence of equivalent reactive quartz content on expansion due to alkali silica reaction. Cement and Concrete Research (41): 99–104.

Andrade, T, Rego Silva, JJ, Martins da Silva, C, Hasparyk, NP (2008): History of some AAR in the Recife region of Brazil. In: Broekmans, MATM, Wigum, BJ (editors), Proceedings of the 13th International Conference on Alkali-Aggregate Reaction in Concrete (ICAAR), Trondheim, Norway: 709–722.

ASTM C294 (2011): Standard descriptive nomenclature for constituents of natural mineral aggregates. American Society for Testing and Materials, West Conshoshocken/PA: pp 10.

ASTM C295 (2012): Standard guide for petrographic examination of aggregates for concrete. American Society for Testing and Materials, West Conshoshocken/PA: Concrete and aggregates: pp 8.

Batic, O, Maiza, P, Sota, J (1994): Alkali silica reaction in basaltic rocks NBRI method. Cement and Concrete Research (24/7): 1317–1326.

Blight, GE, Alexander, MG (2011): Alkali-aggregate reaction and structural damage to concrete. Engineering Assessment, Repair and Management. CRC Press. pp 250.

Brodie, KH, Fettes, D, Harte, B, Schmidt, R (2002): A systematic nomenclature for metamorphic rocks. 3. Structural terms including fault-rock terms. Recommendations by the IUGS Subcommission on the systematics of metamorphic rocks. (free accessible from www.bgs.ac.uk/SCMR).

Broekmans, MATM, Fernandes, I, Nixon, P (2009): A global petrographic atlas of alkali-silica reactive rock types: a brief review. In: Middendorf, B, Just, A, Klein, D, Glaubitt, A, Simon, J (editors), Proceedings of the 12th Euroseminar on Microscopy Applied to Building Materials, 15–19 September, Dortmund: 39–50.

BS 7943 (1999): Guide to the interpretation of petrographical examinations for alkali-silica reactivity. British Standards Institution, London: pp 20.

BS 812: Part 104 (1994): Testing aggregates. Part 104: Methods for qualitative and quantitative petrographic examination of aggregates, British Standards Institution, London: pp 19.

CSA A23.2-15A (2009): Petrographic examination of aggregates. Canadian Standards Association, Mississauga, Ontario: 297–328.

CUR Recommendation 89 (2008): Measures to prevent concrete damage by the alkali-silica reaction. (In Dutch) 2nd revised edition. Centre for Civil Engineering Research and Codes, Gouda, the Netherlands.

Fettes, D, Desmons, J (editors) (2007): Metamorphic Rocks: A Classification and Glossary of Terms. Recommendations of the International Union of Geological Sciences Subcommission on the Systematics of Metamorphic Rocks. British Geological Survey. ISBN-13: 9780521868105.

French, WJ (1991): Concrete petrography: a review. Quarterly Journal of Engineering Geology (24): 17–48.

Grattan-Bellew, PE (1992): Microcrystalline quartz, undulatory extinction and the alkali-silica reaction. In: Poole, AB (editor): Proceedings of the 9th International Conference on Alkali-Aggregate Reaction in Concrete (ICAAR), London, UK: 383–394.
Grattan-Bellew, P, Mitchell, LB (2006): Quantitative petrographic analysis of concrete - the damage rating index (DRI) method, a review. In: Fournier B (editor), Marc-André Bérubé Symposium on alkali-aggregate reactivity in concrete. 8th CANMET/ACI International Conference on Recent Advances in Concrete Technology, Montréal/CA: 321–334.
Hallsworth, CR, Knox, RWO'B (1999): BGS Rock Classification Scheme. Volume 3. Classification of sediments and sedimentary rocks. British Geological Survey Research Report, 2nd Edition, RR 99-03.
Hasparyk, NP, Lopes, ANM, Cavalcanti, AJCT, Silveira, JFA (2004): Deterioration index and properties of concretes from Paulo Afonso Power Plants-Brazil-affected by the alkali-aggregate reaction. In: Mingshu, T, Min, D (editors), Proceedings of the 12th International Conference on Alkali-Aggregate Reaction in Concrete (ICAAR), International Academic Publishers, Beijing, China: 898–906.
Hasparyk, NP, Monteiro, PJM, Molin, DCCD, (2008): AAR in Furnas dam, Brazil, residual expansions and the effect of lithium. In: Broekmans, MATM, Wigum, BJ (editors), Proceedings of the 13th International Conference on Alkali-Aggregate Reaction in Concrete (ICAAR), Trondheim, Norway: 87–96.
Humphries, DW (1992): The preparation of thin-sections of rocks, minerals and ceramics. Royal Microscopical Society, Oxford Science Publications, Microscopy Handbooks (24): pp 83.
Hünger, KJ (2005): Zum Reaktionsmechanismus präkambrischer Grauwacken aus der Lausitz bei ihrer Verwendung als Gesteinskörnung im Beton. Habilitationsschrift, Brandenburgischen Technische Universität Cottbus, Shaker Verlag, Aachen/DE: pp 235.
Le Maitre, RW, Streckeisen, A, Zanettin, B, Le Bas, MJ, Bonin, P, Bateman, P, Bellieni, G, Dudek, A, Efremova, S, Keller, J, Lameyre, J, Sabine, PA, Schmidt, R, Sørensen, H, Woolley, AR (2005): A classification of igneous rocks and glossary of terms. Recommendations of the International Union of Geological Sciences, Subcommission on the Systematics of Igneous Rocks. 2nd edition. Blackwell Scientific, Oxford: pp 256.
Leemann, A, Merz, C (2012a): An attempt to validate the concrete performance test with the degree of AAR-induced damage observed in concrete structures. In: Drimalas, T, Ideker, JH, Fournier, B (editors), Proceedings of the 14th International Conference on Alkali-Aggregate Reactions in Concrete (ICAAR), Austin, USA, pp 10.
Leemann, A, Merz, C (2012b): Comparison between AAR-induced expansion determined with an ultra-accelerated microbar test and a concrete performance test. In: Drimalas, T, Ideker, JH, Fournier, B (editors), Proceedings of the 14th International Conference on Alkali-Aggregate Reactions in Concrete (ICAAR), Austin, USA, pp 10.
Lorenzi, G, Jensen, J, Wigum, B, Sibbick, R, Haugen, M, Guédon, S, Åkesson, U (2006): Petrographic atlas of the potentially alkali-reactive rocks in Europe. FP5—Partner-project-GRD1-CT-2001-40103. Geological Survey of Belgium Professional Paper 2006/1-N°302: pp 63.
MacKenzie, WS, Donaldson, CH, Guilford, C (1982): Atlas of igneous rocks and their textures. Essex, England: Longman Group Limited: pp 148.
NCA (2004): Alkali Aggregate Reactions in Concrete. Test Methods and Requirements to Test Laboratories, Norwegian Concrete Association, Publication No 3 (in Norwegian).
Nixon, PJ, Sims, I (2016): RILEM recommendations for the prevention of damage by alkali-aggregate reactions in new concrete structures, State-of-the-Art Report of the RILEM Technical Committee 219-ACS, RILEM State-of-the-Art Reports, Volume 17, Springer (Dordrecht, The Netherlands) RILEM (Paris).
Oberholster, RE (1996): Case studies of the practical and economical impact of alkali-silica reaction in South Africa. In: Shayan, A (editor), Proceedings of the 10th International Conference on Alkali-Aggregate Reaction (ICAAR), Melbourne, Australia: 123–132.
Oberholster, RE (2009): Alkali-silica reaction. In: Owens, G (editor), Fulton's Concrete Technology, 9th edition, Midrand, South Africa, Cement and Concrete Institute: 189–218.
RILEM AAR-1 (2003): Detection of potential alkali-reactivity of aggregates—petrographic method. Materials and Structures (36/7): 480–496.
RILEM (2016): RILEM Recommended Test Method: AAR-1.1 – Detection of potential alkali-reactivity – Part 1: Petrographic examination method. In: Nixon, PJ, Sims, I (editors): RILEM recommendations for the prevention of damage by alkali-aggregate reactions in new concrete structures. RILEM State-of-the-art Report (17): 35–60.
Rogers, C, MacDonald, C-A (2012): The geology, properties and field performance of alkali-aggregate reactive Spratt, Sudbury and Pittsburg aggregates distributed by the Ontario Ministry of Transportation. In: Drimalas, T, Ideker, JH, Fournier, B (editors), Proceedings of the 14th International Conference on Alkali-Aggregate Reactions in Concrete (ICAAR), Austin, USA.
Sanchez, L, Salva, P, Fournier, B, Jolin, M, Pouliot, N, Hovington, A (2012): Evaluation of damage in the concrete elements of the Viaduct "Robert-Bourassa-Charest" after nearly 50 years in service. In: Drimalas, T, Ideker JH, Fournier, B (editors), Proceedings of the 14th International Conference on Alkali-Aggregate Reactivity in Concrete (ICAAR), Austin, USA, 081511-FOUR-01. pp 10.
Silveira, JFA, Degaspare, JC, Cavalcanti, AJCT, (2000): Diagnosis of the cause of the progressive concrete swelling at the Paulo Afonso I, II and III underground power stations. In: Bérubé, MA, Fournier, B, Durand, B (editors), Proceedings of the 11th International Conference on Alkali-Aggregate Reaction in Concrete (ICAAR), Québec: 1393–1402.
St John, DA, Poole, AB, Sims, I (1998): Concrete petrography – A handbook of investigative techniques, Arnold, U.K, pp 474; (now Poole, AB, Sims, I (2016): Concrete petrography, a handbook of investigative techniques. 2nd edition. CRC Press, London/UK: pp 802).
Vernon, RH (2011): A practical guide to rock microstructure. Cambridge University Press. pp 594.
Wigum, BJ (2000): "Normin2000"—A Norwegian AAR research program. In: Bérubé, MA, Fournier, B, Durand, B (editors), Proceedings of the 11th International Conference on Alkali-Aggregate Reaction in Concrete (ICAAR), Québec, CA: 523–531.

Igneous Rocks

2

Isabel Fernandes, Helena Martins, Maria dos Anjos Ribeiro, Fernando Noronha, Maarten A.T.M. Broekmans, and Ian Sims

Igneous rocks form from cooling and solidification of molten rock, either from magma within the Earth's crust (plutonic rocks), or from lava extruded on to the Earth's surface in the atmosphere or underwater (volcanic rocks).

Igneous rocks can be divided into two broad textural classes: 1. coarsely crystalline rocks, which are intrusive (plutonic) and therefore cooled slowly, and 2. finely crystalline to glassy/amorphous rocks, which are extrusive (volcanic) and cooled rapidly. In between these two classes hypabyssal rocks present an intermediate size of the crystals. Within each of these textural classes, the rocks are subdivided according to their composition. General compositional classes of igneous rocks are felsic, intermediate, mafic and ultramafic, in decreasing silica and increasing iron and magnesium contents.

Igneous rocks of different types are widely used as aggregates for concrete. The most common are of volcanic origin, such as basalts, andesites and rhyolites, as well as plutonic rocks of "granitic" composition. However, the features related to the potential reactivity differ from one rock type to another and there seems to be an important dependence on the regional geologic history of the rock.

The potential alkali-reactivity of volcanic rocks has been investigated in many different locations around the world including Argentina, Australia, Brazil, China, Iceland, Japan, New Zealand and Turkey (e.g. Katayama and Kaneshige 1986; Katayama et al. 1989; Batic et al. 1994; Freitag and St. John 1996; Goguel 1996; Katayama 1997; Guðmundsson and Ólafsson 1999; Wakizaka 2000; Marfil and Maiza 2001, 2006; Korkanç and Tugrul 2004; Shayan 2004; Çopuroğlu et al. 2009). Most of these rocks do not contain mineral phases of pure silica but the presence of rhyolitic (= rich in Si, poor in Fe, Mg) glass in the matrix is generally accepted as the cause for the alkali-reactivity potential of the rock. Volcanic rocks contain, or may contain, secondary minerals that are alkali-reactive and/or release alkalis into the concrete pore solution (Alexander and Mindess 2005). Also, chalcedony, cristobalite, devitrified glass and tridymite are potentially reactive forms of silica (Blight and Alexander 2011). However, the fine-grained nature of the volcanic rocks and the presence of glassy matrix make the identification of reactive forms of silica quite challenging.

Causes of the potential alkali-reactivity of coarse-grained plutonic rocks, like granite, diorite and related rock types, remain a matter of discussion and these rocks show great variation in performance, as much in field structures as in expansion testing. One of the factors that needs clarification and seems to divide geologists is the definition of limits between granite (without preferential orientation of the crystals), deformed granitic rock with preferred mineral orientation texture and granitic gneiss with layering and banding. For granite, the potential alkali-reactivity is primarily attributed to the content of microcrystalline quartz (≤100 μm) and/or lattice strain (e.g. Gogte 1973; Grattan-Bellew 1992; Kerrick and Hooton 1992; Shayan 1993; Wenk et al. 2008). However, microcracks in the aggregate particles also provide easy access to quartz grains in the interior, as suggested by Velasco-Torres et al. (2010).

I. Fernandes (✉)
ICT, Institute of Earth Sciences and Department of Geology, Faculty of Sciences, University of Lisbon, Lisbon, Portugal
e-mail: mifernandes@fc.ul.pt

H. Martins · M.A. Ribeiro · F. Noronha
ICT, Institute of Earth Sciences and DGAOT, Faculty of Sciences, University of Porto, Porto, Portugal
e-mail: hbrites@fc.up.pt

M.A. Ribeiro
e-mail: maribeir@fc.up.pt

F. Noronha
e-mail: fmnoronh@fc.up.pt

M.A.T.M. Broekmans
Department of Mineral Resources, Geological Survey of Norway - NGU, Trondheim, Norway
e-mail: maarten.broekmans@ngu.no

I. Sims
RSK Environment Ltd, Hemel Hempstead, UK
e-mail: ian@simsdoc.com; isims@rsk.co.uk

I. Fernandes et al. (eds.), *Petrographic Atlas: Characterisation of Aggregates Regarding Potential Reactivity to Alkalis*, DOI 10.1007/978-94-017-7383-6_2

The designation of the rock also depends on local geological context and the limits between granite and granitic gneiss are frequently diffuse and dependent on local experience and nomenclature. As such, rocks of granitic composition, in this atlas designated as "granite", show quite variable texture. This designation could be different if field information were available and taken into account in establishing a rock name.

The nomenclature applied here follows the classifications developed by Le Maitre et al. (2005), in which both the mineral composition and the texture of the rocks are used. Due to the objective of this study, the main division is made by grain size, therefore dividing the igneous rocks into plutonic (coarser-grained) and volcanic (fine to very-fine grained).

In the descriptions and in the photomicrographs some metamorphic features can be found. However, whenever the original texture of the rock maintains igneous characteristics, the sample was included in this chapter. In some ambiguous cases (owing to not having information about the geologic context of the rock) the reference to metamorphic features was included in the description of the sample.

References

Alexander, M, Mindess, S (2005): Aggregates in concrete. Taylor & Francis, London, UK. Modern Concrete Technology (13): pp 435.

Batic, O, Maiza, P, Sota, J (1994): Alkali silica reaction in basaltic rocks NBRI method. Cement and Concrete Research (24/7): 1317–1326.

Blight, GE, Alexander, MG (2011): Alkali-aggregate reaction and structural damage to concrete. Engineering Assessment, Repair and Management. CRC Press. pp 250.

Çopuroğlu, O, Andiç-Çakir, Ö, Broekmans, MATM, Kühnel, RA (2009): Mineralogy, geochemistry and expansion testing of an alkali-reactive basalt from Western Anatolia, Turkey. Materials Characterization (60/7), Special Issue (35): 756–766.

Freitag, SA, St John, DA (1996): Alkali aggregate reaction in existing structures – what can it tell us? In: Shayan, A (editor), Proceedings of the 10th International Conference on Alkali-Aggregate Reaction (ICAAR), Melbourne, Australia: 183–190.

Gogte, BS (1973): An evaluation of some common Indian rocks with special reference to alkali-aggregate reactions. Engineering Geology (7): 135–153.

Goguel, R (1996): Selective dissolution techniques in AAR investigation: application to an example of failed concrete. In: Shayan, A (editor), Proceedings of the 10th International Conference on Alkali-Aggregate Reaction (ICAAR), Melbourne, Australia: 783–790.

Grattan-Bellew, PE (1992): Microcrystalline quartz, undulatory extinction and the alkali-silica reaction. In: Poole, AB (editor): Proceedings of the 9th International Conference on Alkali-Aggregate Reaction in Concrete (ICAAR), London, UK: 383–394.

Guðmundsson, G, Ólafsson, H (1999): Alkali-silica reactions and silica fume, 20 years of experience in Iceland. Cement and Concrete Research (29): 1289–1297.

Katayama, T (1997): Petrography of alkali-aggregate reactions in concrete – reactive minerals and reaction products. In: Nishibayashi, S, Kawamura, M (editors), East Asia Alkali-Aggregate Reaction Seminar, Tottori: A45–A59.

Katayama, T, Kaneshige, Y (1986): Diagenetic changes in potential alkali aggregate reactivity of volcanic rocks in Japan-A geological interpretation, In: Grattan-Bellew, PE (editor): Proceedings of the 7th International Conference on Alkali-Aggregate Reaction in Concrete (ICAAR), Ottawa, Canada: 489–495.

Katayama, T, St John, DA, Futagawa, T (1989): The petrographic comparison of rocks from Japan and New Zealand-Potential reactivity related to interstitial glass and silica minerals. In: Okada, K, Nishibayashi, S, Kawamura, M (editors), Proceedings of the 8th International Conference on Alkali-Aggregate Reaction in Concrete (ICAAR), Kyoto, Japan: 537–542.

Kerrick, D, Hooton, R (1992): ASR of concrete aggregate quarried from a fault zone: results and petrographic interpretation of accelerated mortar bar tests. Cement and Concrete Research (22): 949–960.

Korkanç, M, Tugrul, A (2004): Evaluation of selected basalts from Nigde, Turkey, as source of concrete aggregate. Engineering Geology (75/3–4): 291–307.

Le Maitre, RW, Streckeisen, A, Zanettin, B, Le Bas, MJ, Bonin, P, Bateman, P, Bellieni, G, Dudek, A, Efremova, S, Keller, J, Lameyre, J, Sabine, PA, Schmidt, R, Sørensen, H, Woolley, AR (2005): A classification of igneous rocks and glossary of terms. Recommendations of the International Union of Geological Sciences, Subcommission on the Systematics of Igneous Rocks. 2nd edition. Blackwell Scientific, Oxford: pp 256.

Marfil, S, Maiza, P (2001): Basaltic rocks: their use as concrete aggregates. In: Väisanen, P, Uusinoka, R (editors), Aggregate 2001, Environment and Economy. Finland (1): 203–206.

Marfil, SA, Maiza, P (2006): Assessment of the potential alkali reactivity of rhyolitic rocks from Argentina, IAEG2006 Paper number 769, The Geological Society of London: 1–6.

Shayan, A (1993): Alkali reactivity of deformed granitic rocks: a case study. Cement and Concrete Research (23): 1229–1236.

Shayan, A (2004): Alkali-aggregate reaction and basalt aggregates. In: Tang, M, Deng, M (editors), Proceedings of the 12th International Conference on Alkali-Aggregate Reactions in Concrete (ICAAR), Beijing, China: 1130–1135.

Velasco-Torres, A, Alaejos, P, Soriano, J (2010): Comparative study of the alkali-silica reaction (ASR) in granitic aggregates. Estudios Geológicos (66/1): 105–114.

Wakizaka, Y (2000): Alkali-silica reactivity of Japanese rocks. Engineering Geology (56): 211–221.

Wenk, H-R, Monteiro, PJM, Shomglin, K (2008): Relationship between aggregate microstructure and mortar expansion. A case study of deformed granitic rocks from Santa Rosa mylonite zone. Journal of Materials and Science (43): 1278–1285.

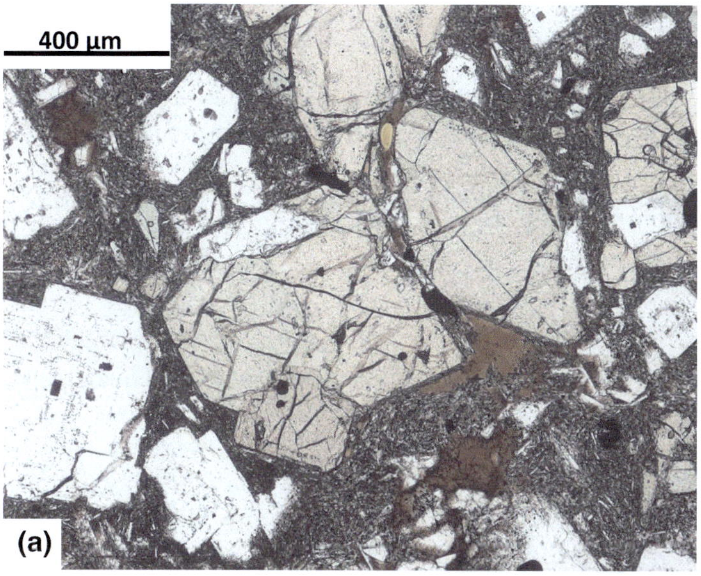

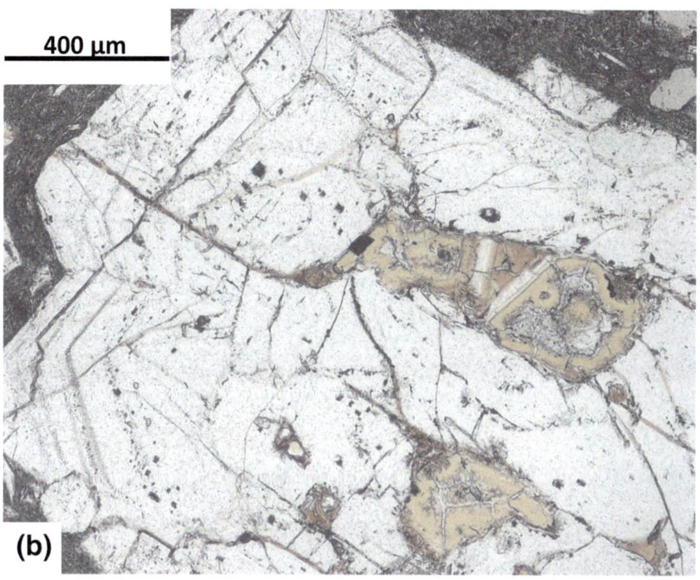

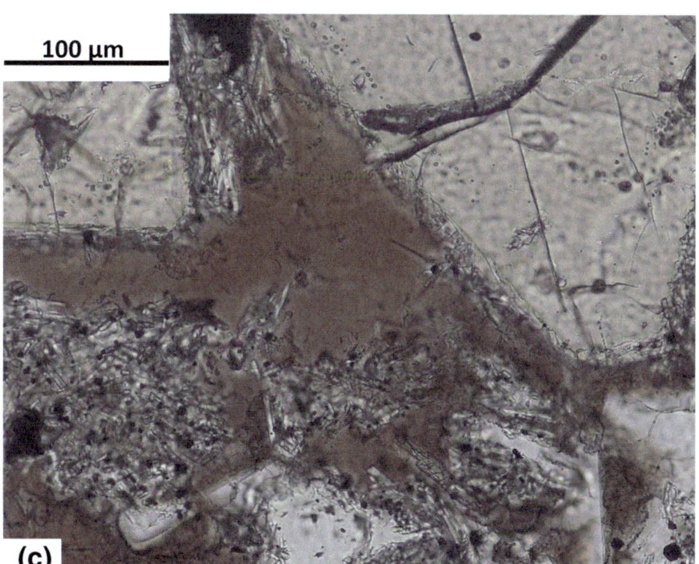

Andesite (JP)

Moderately altered andesite showing a porphyritic texture with hypocrystalline groundmass (a, PPL). Euhedral phenocrysts of pyroxene and plagioclase are abundant and display weak alteration. Pyroxene (very light brown) is locally altered to smectite (b, PPL). The groundmass is composed of volcanic glass and tiny laths of plagioclase, pyroxene (pigeonite-augite), olivine and opaque minerals. Volcanic glass (brown) is observed in interstitial spaces (c, PPL), and is also seen to form inclusions in zoned plagioclase phenocrysts.

Reactivity potential attributed to:
Rhyolitic volcanic glass (according to literature).

Aggregate particle in a sample of concrete (Japan).

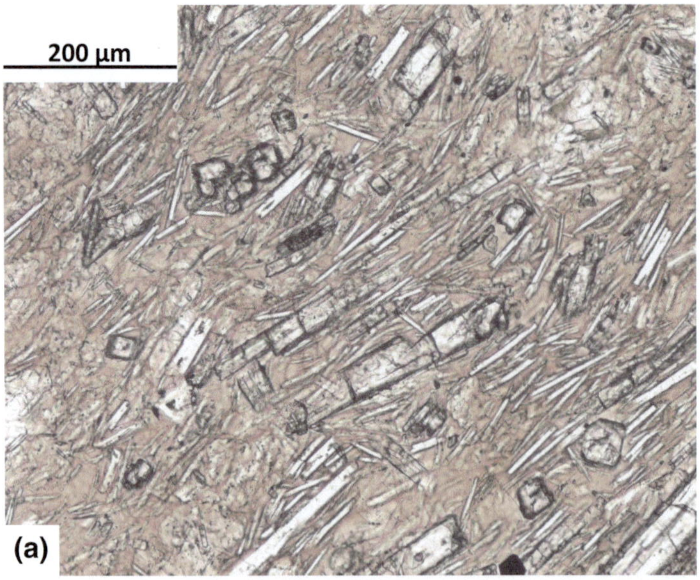

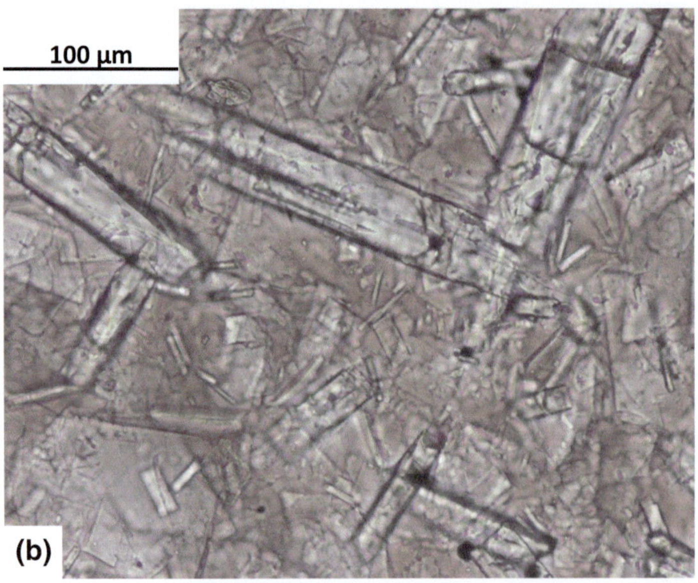

Andesite (JP)

Glassy aphyric bronzite andesite composed of main plagioclase, minor bronzite pyroxene, and traces of opaque minerals embedded in amorphous volcanic glass (a, PPL). Tiny laths of plagioclase and microphenocrysts of orthopyroxene (bronzite) float in the amorphous clove-brown volcanic glass groundmass. Subhedral crystals of pyroxene and plagioclase exhibit a preferred fluidal orientation. The larger prismatic laths represent bronzite (b, PPL). The rock is very little altered and the groundmass may contain scarce cristobalite and tridymite. The isotropic volcanic glass is the main component of the groundmass (c, XPL). It might correspond to rhyolitic glass.

Reactivity potential attributed to:
Rhyolitic volcanic glass (according to literature).

Aggregate particle in a sample of concrete from a railway structure. Also proved to be reactive in many types of structures (Japan).

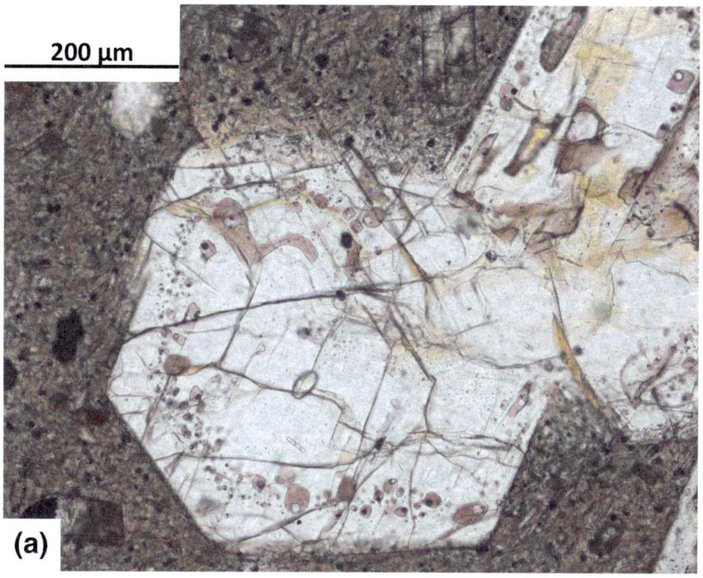

(a)

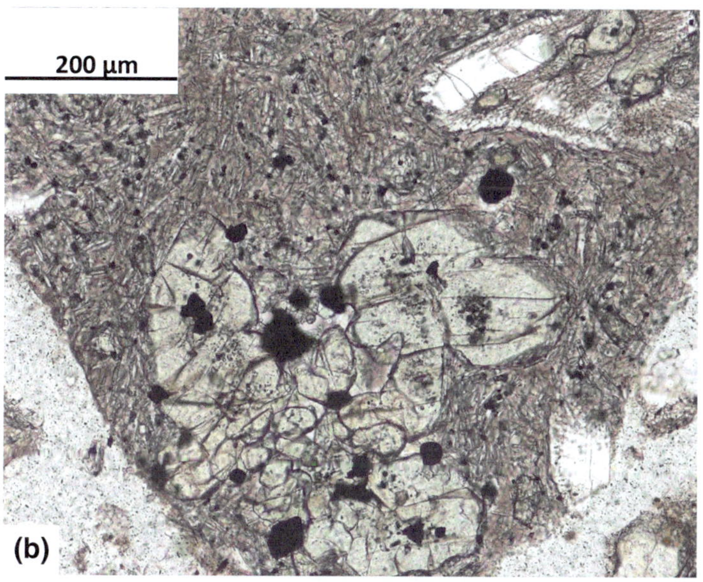

(b)

(c)

Andesite (TR)

Slightly altered andesite with porphyritic texture composed of plagioclase, pyroxene (augite, bronzite and hypersthene) olivine and trace amounts of opaque minerals (magnetite). Crystals of olivine have been replaced by carbonate minerals. The phenocrysts of pyroxene (augite, bronzite) are mostly euhedral, show fractures and abundant melt inclusions, in some crystals arranged in zones (a, PPL). The groundmass is hyalopilitic and contains small grains of plagioclase, augite, hypersthene and abundant volcanic glass (brown) (b, PPL). The glassy matrix is more abundant than the phenocrysts content and is mainly isotropic (c, XPL).

Reactivity potential attributed to:
Volcanic glass (dacitic, according to literature) in the matrix.

Sample of rock used as aggregate in concrete prisms (Turkey).

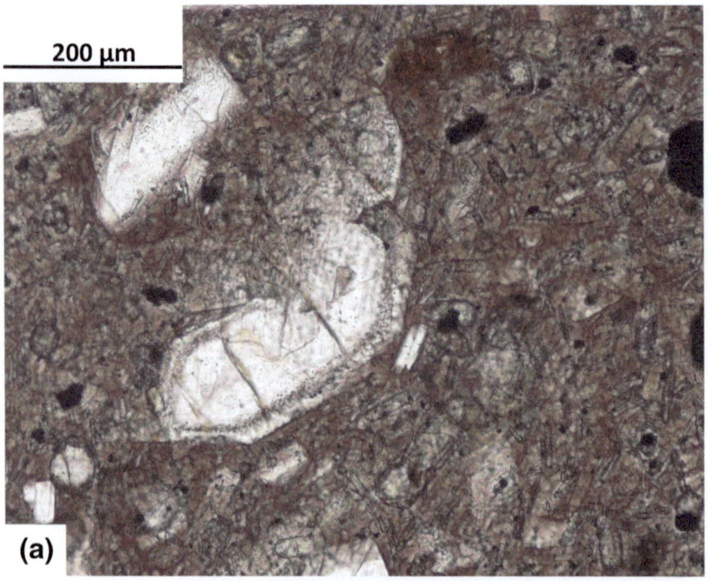

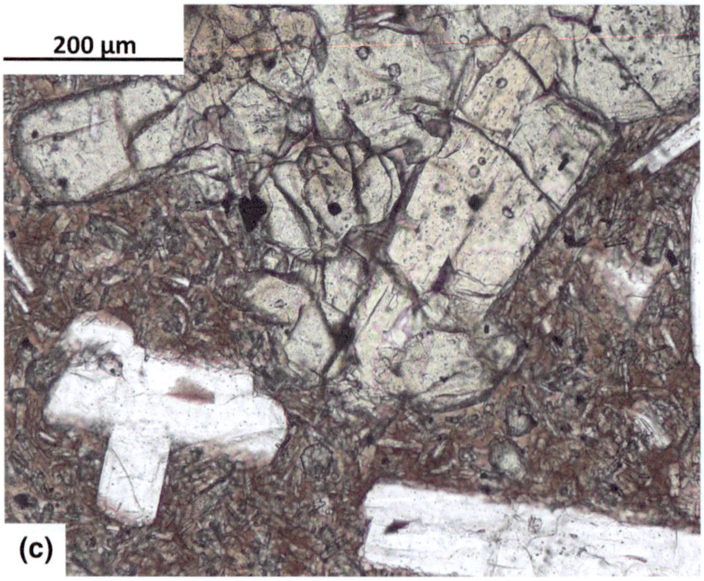

Andesite (TR)

Slightly altered andesite showing porphyritic texture and hyalopilitic groundmass (a, PPL). Phenocrysts consist of euhedral plagioclase, pyroxene (augite and bronzite). The rock is coloured brown from the volcanic glass matrix, also occurring as inclusions in larger phenocrysts (in zones, in addition to oscillatory zoning – b XPL). The matrix contains tiny crystals of acicular plagioclase, pyroxene (hypersthene, bronzite and augite) and opaque minerals (c, PPL). The glassy matrix is more abundant than the phenocrysts content.

Reactivity potential attributed to:
Volcanic glass (dacitic, according to literature) in the matrix.

Sample of rock used as aggregate in concrete prisms (Turkey).

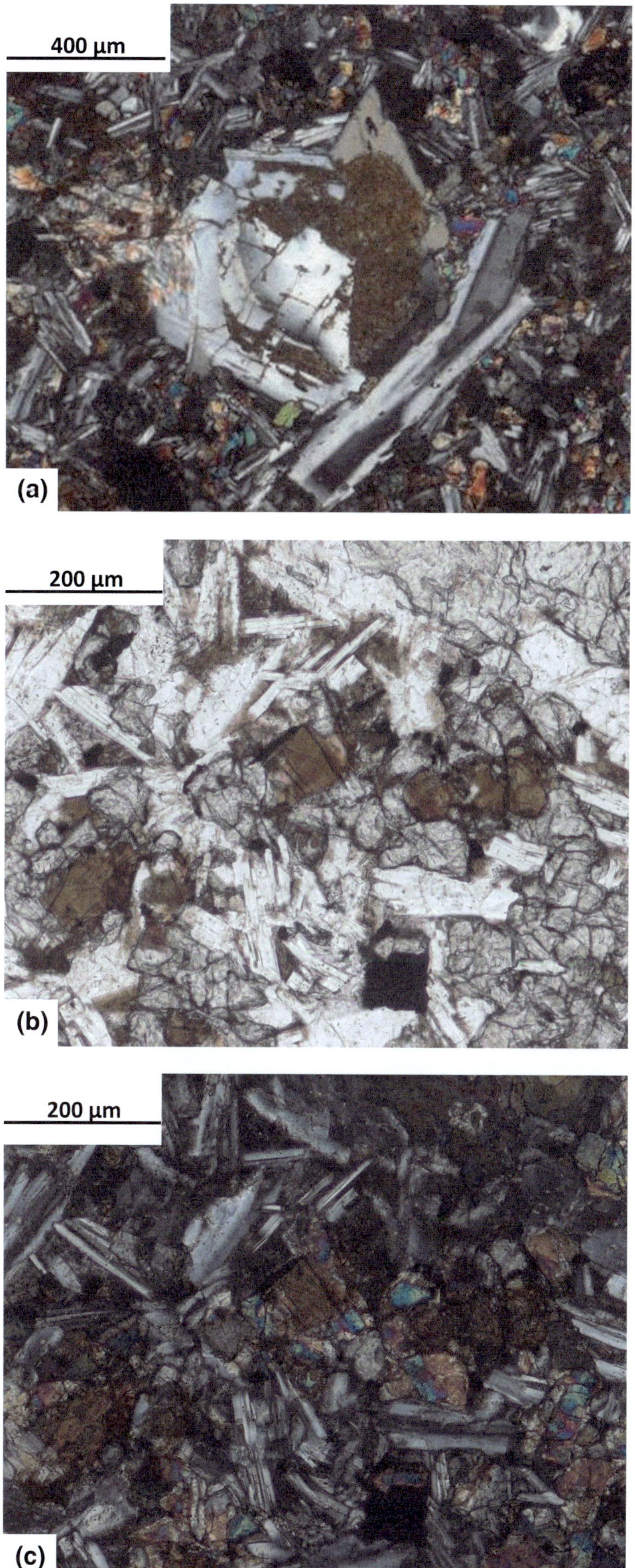

Basalt (AR)

Olivine basalt with interstitial texture composed of plagioclase, pyroxene (titanaugite), olivine, trace amounts of opaque minerals (hematite, rutile), as well as sideromelane, montmorillonite, chlorite and cryptocrystalline silica of secondary origin. This fine-grained rock contains small phenocrysts of plagioclase, olivine and pyroxene in clusters. The subhedral andesitic plagioclase is zoned and slightly altered in its interior (a, XPL). The titaneous augite lost the iron thus forming oxides and hydroxides. The augite crystals are located between the feldspar crystals and also in the groundmass (b, PPL). The voids between phenocrysts are filled with volcanic glass, commonly altered to montmorillonite, chlorite, iron and titanium hydroxides and cryptocrystalline silica (c, XPL). The groundmass is mainly composed of plagioclase and devitrified volcanic glass. Relicts ofpalagonite are found in some sectors, as well as montmorillonite (brownish) and, less commonly, chlorite, serpentine, apatite and iron oxides.

Reactivity potential attributed to:
Cryptocrystalline silica and volcanic glass (composition to be confirmed by micro-analysis).

Sample of rock used as aggregate for concrete dams and pavements (Argentina).

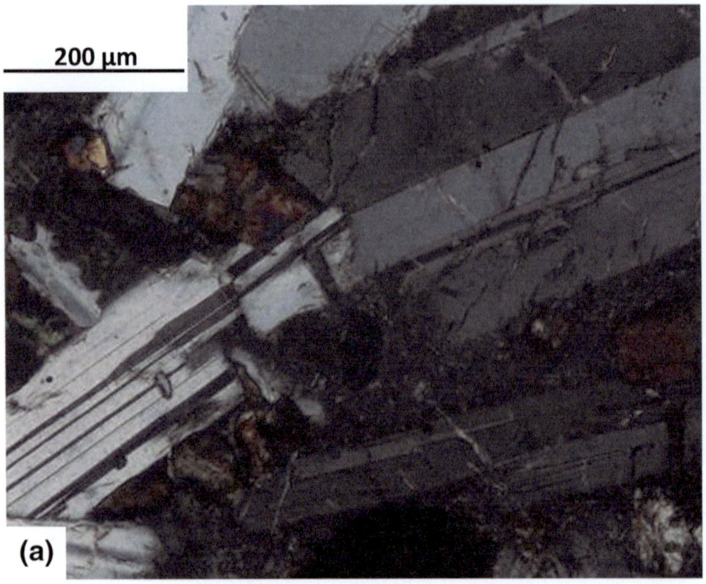

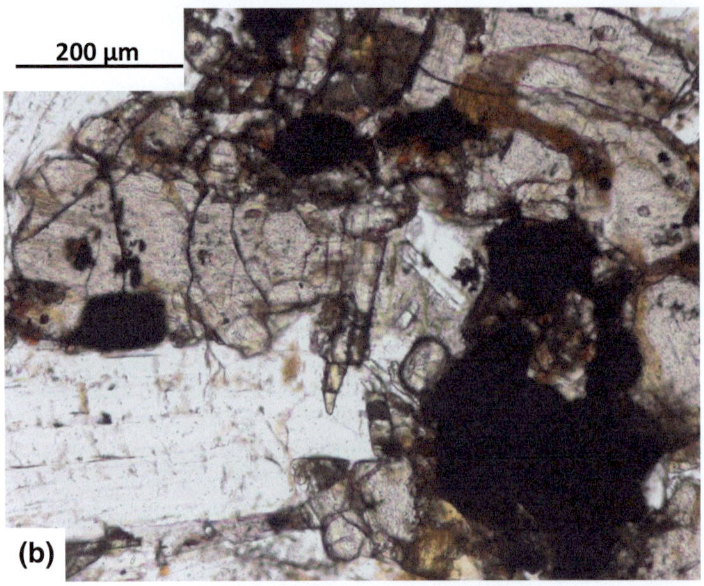

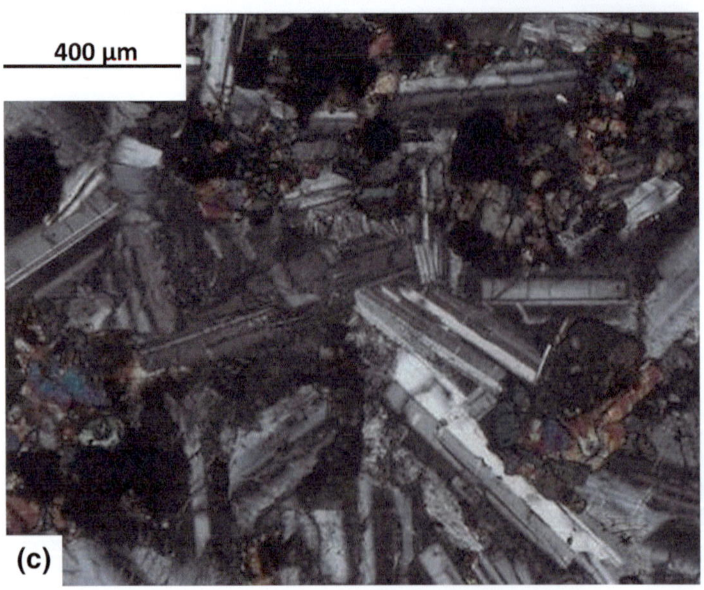

Basalt (AR)

Moderately altered basalt showing intersertal texture and composed of plagioclase, pyroxene (moderately to intensely altered), montmorillonite, volcanic glass (devitrified and moderately altered) and opaque minerals. Plagioclase is the main constituent of the rock and occurs as large laths (a, XPL), with interstitial pyroxene (pinkish-greenish pleochroic), opaque minerals and volcanic glass (brown) (b, PPL). Pyroxene phenocrysts are rare, and are altered to clay minerals. Under XPL, the volcanic glass is isotropic (c, XPL). Cracks are abundant in the rock.

Reactivity potential attributed to:
Interstitial volcanic glass (composition to be confirmed by micro-analysis).

Sample of rock used as aggregate for concrete dams and pavements (Argentina).

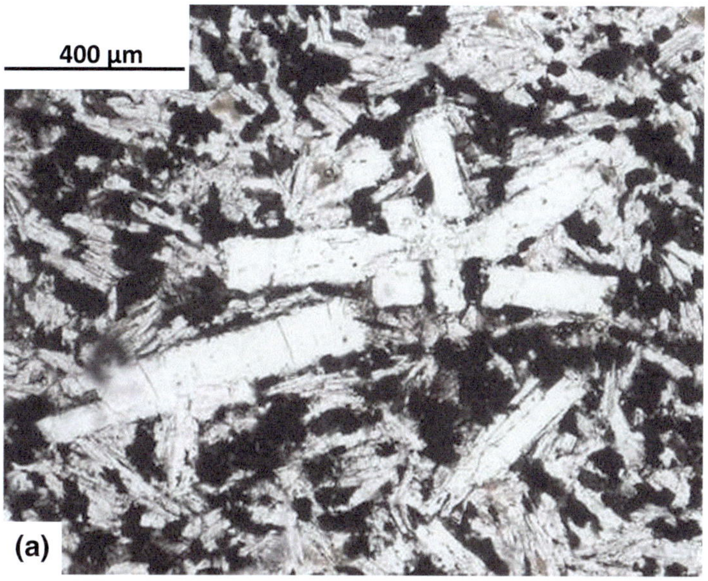

Basalt (CN)

Basalt composed of plagioclase laths, occurring both as phenocrysts and in the fine-grained matrix, as well as opaque minerals, iron oxides, smectite and volcanic glass (a, PPL). This dense basalt is slightly altered and shows intersertal texture. It is mainly composed of plagioclase laths in which the phenocrysts are bunched together (b, XPL). Opaque minerals are the second component in abundance. Volcanic glass is scarce and occurs in the interstitial spaces (brownish) (c, PPL).

Reactivity potential attributed to:
Volcanic glass in the matrix (composition to be confirmed by micro-analysis).

Sample of rock classified as potentially reactive in laboratory tests (China).

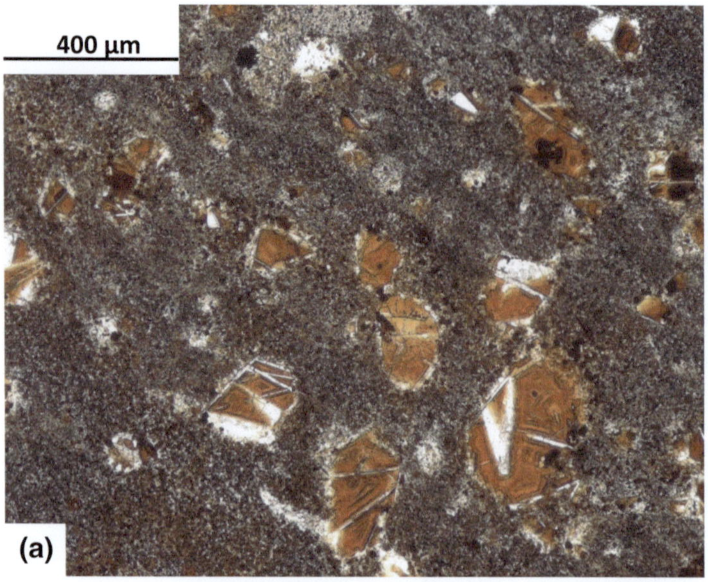

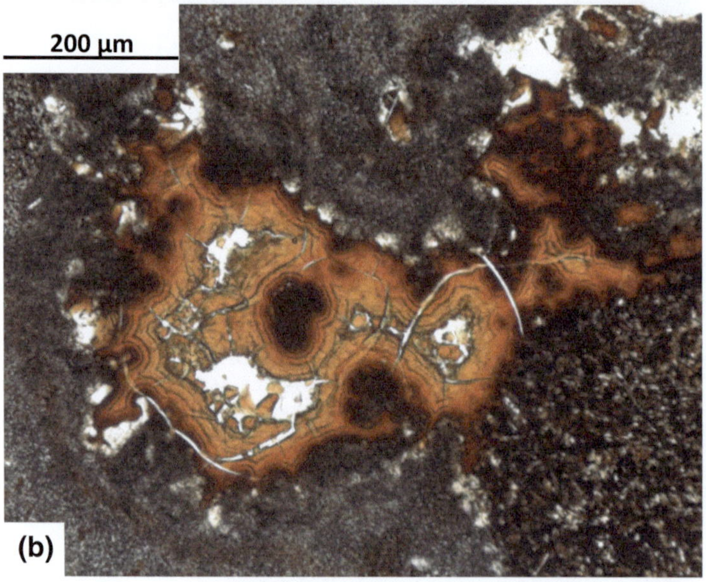

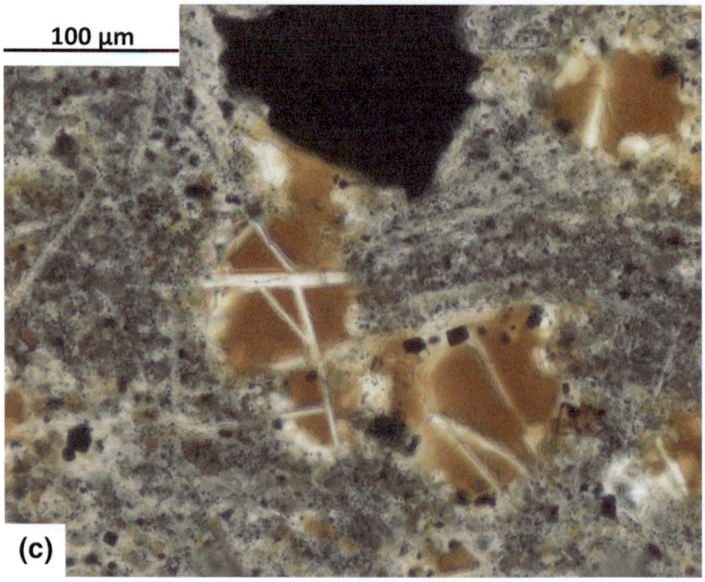

Dacite (IS)

Rock of intermediate composition showing porphyritic texture (a, PPL) and intense alteration. The rock is composed of phenocrysts of clynopyroxene altered to smectite minerals (brownish orange) and quartz (colourless) (b, PPL). Opaque minerals are usually associated with the clynopyroxene. Silica is present as tridymite (wedged shape crystals), cristobalite and quartz (lighter colour patches in the groundmass). Devitrified glass with cryptocrystalline texture is abundant in the groundmass (c, PPL).

Reactivity potential attributed to:
Cryptocrystalline quartz in the groundmass, tridymite and cristobalite (according to literature).

Sample of rock classified as potentially reactive (Iceland).

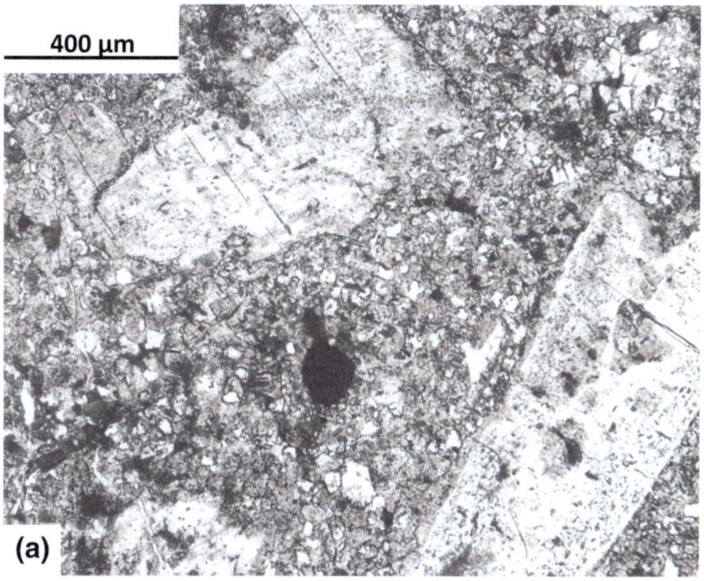

Felsic porphyry (AO)

Felsic igneous rock of porphyritic texture mainly composed of quartz, plagioclase, K-feldspar, amphibole (green hornblende) and pyroxene. The phenocrysts are embedded in a very fine-grained felsic groundmass containing quartz and feldspar (a, XPL), sericite, chlorite (vermicular), epidote, apatite, monazite and opaque minerals. The quartz phenocrysts show irregular boundaries in the contact with the matrix (b, XPL). Hydrothermal alteration is intense and caused the formation of abundant sericite in the plagioclase and the replacement of pyroxene by amphibole and chlorite. There are abundant intergrowths of quartz and feldspar and the matrix dominantly exhibit granophyric texture (c, XPL).

Reactivity potential attributed to:
Microcrystalline quartz in interstitial spaces.

Aggregate particle in a sample of concrete from a bridge (Angola).

Felsic porphyry (DE)

Felsic igneous rock with porphyritic texture and composed of quartz, K-feldspar, plagioclase, biotite, chlorite, muscovite, and traces of zircon, sericite, apatite and opaque minerals. The rock contains large phenocrysts of quartz, K-feldspar and biotite, floating in a finely crystalline matrix of quartz and feldspar (a, XPL) also containing small flakes of muscovite and chloritised biotite. The well-developed quartz crystals show embayments by corrosion gulfs (b, XPL). More often, quartz forms small crystals and abundant intergrowths with the alkali-feldspar in the microcrystalline matrix (c, XPL).

Reactivity potential attributed to:
Microcrystalline quartz in unstable intergrowths in the groundmass.

Aggregate particle in a sample of concrete with ASR from highway pavement (Germany).

Felsic porphyry (SE)

Rock of granitic composition showing porphyritic texture and composed of main quartz, K-feldspar, minor plagioclase, and traces of chlorite and iron hydroxides. Quartz and feldspar phenocrysts are dispersed in a very fine-grained, holocrystalline matrix composed of quartz, feldspar and some chlorite (a, XPL). Locally, the quartz phenocryst edges are corroded due to subgraining and recrystallization (b, XPL). The irregular boundaries of the phenocrysts are marked by agglomerates of small crystals of quartz (c, XPL).

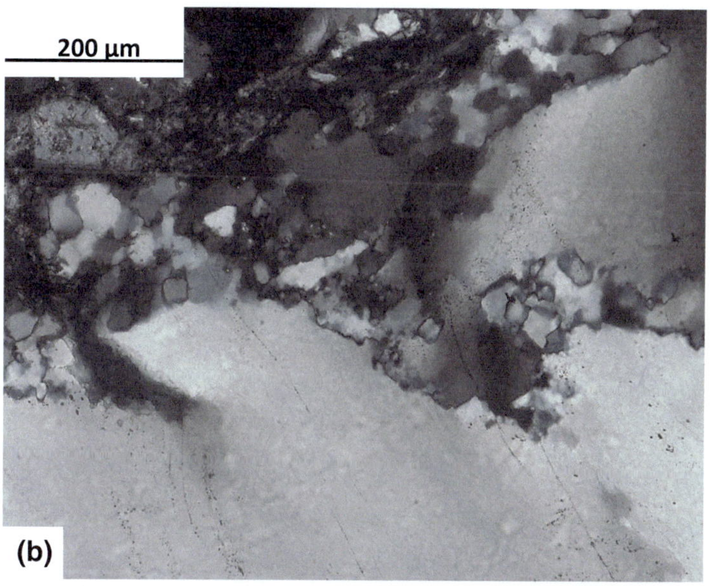

Reactivity potential attributed to:
Microcrystalline quartz in the groundmass.
Strained phenocrysts with sutured boundaries.

Aggregate particle in a sample from a concrete prism (Sweden).

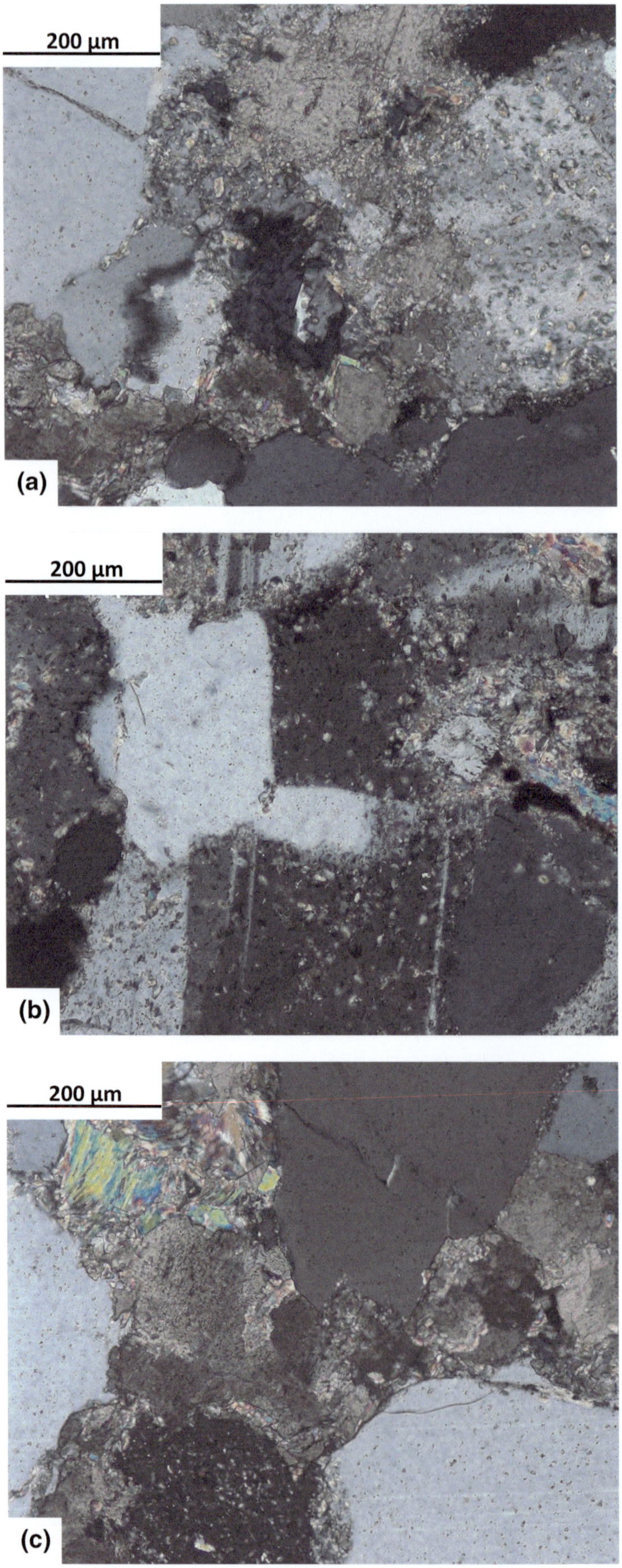

Felsic porphyry (ZA)

Igneous rock of granitic composition containing quartz, plagioclase, minor sericite, chlorite, muscovite, epidote, calcite and opaque minerals. The phenocrysts of K-feldspar and quartz are dominant in comparison with the scarce groundmass (a, XPL). The crystals are anhedral to subhedral and show straight to slightly indented boundaries and seldom concave-convex limits. Hydrothermal alteration of the porphyry caused the formation of abundant sericite in the plagioclase. Calcite, chlorite and microcrystalline quartz are found in the interstitial spaces between the larger crystals (c, XPL). Plates of muscovite (secondary) grow perpendicular to the boundaries of quartz and feldspar crystals.

Reactivity potential attributed to:
Interstitial microcrystalline quartz as alteration by-product.

Aggregate particle in concrete from a taxiway (South Africa).

Granite (BR)

Granite composed of quartz, K-feldspar (microcline and orthoclase), plagioclase, biotite, chlorite, epidote, muscovite, allanite and opaque minerals. This deformed biotite granite shows variations in the texture and granularity. There are agglomerations of biotite (greenish) and clots of epidote in an inequigranular texture (a, XPL). Some sectors are dominated by developed crystals of quartz and feldspar with straight borders and showing no deformation while other sectors show fine granularity. The later exhibit microcrystalline quartz and feldspars (b, XPL). The quartz shows some sutured boundaries and there are cracks in the feldspar crystals. K-feldspar is perthitic and some of the intergrowths are bent. Tiny inclusions of quartz are common and there are subgrains (<50 µm) and myrmekites (c, XPL). Chloritised biotite crystals commonly occur in the interstices of the other minerals.

Reactivity potential attributed to:
Strained quartz, recrystallized in some domains. Microcrystalline quartz and abundant myrmekites.

Sample of rock used as aggregate in a dam (Brazil).

Granite (DE)

Medium-grained biotite granite with hypidiomorphic texture (a, XPL) composed of quartz, K-feldspar, plagioclase, biotite and trace amounts of opaque minerals, sericite and zircon. K-feldspar (orthoclase) occurs in large anhedral crystals exhibiting weak alteration (cloudy) and as interstitial material. Plagioclase occurs mainly in the size of the matrix crystals with stronger alteration in the nucleus (b, XPL). Quartz occurs as clear crystals with moderate undulatory extinction (5-15°), as well as microcrystalline material in interstitial spaces (sub-solid recrystallization). The boundaries are straight to slightly curved. There is abundant microcrystalline quartz in myrmekites (c, XPL), in fringes of the plagioclase prismatic crystals showing discontinuous compositional zonation.

Reactivity potential attributed to:
Abundant microcrystalline quartz (myrmekites).

Aggregate particle in a sample of concrete with ASR from a hydraulic structure (Germany).

Granite (ES)

Granite composed of quartz, K-feldspar, plagioclase, muscovite, biotite, chlorite, and traces of zircon, epidote and sericite.
The rock shows hypidiomorphic inequigranular texture. It is slightly to moderately weathered, with strong sericitization of slightly calcium-richer zones in plagioclase (a, XPL). Quartz occurs both in developed anhedral crystals showing slight undulatory extinction (<15°) (b, XPL), and also in small crystals in the interstitial spaces between the larger crystals. Small blobs of quartz are present in feldspar myrmekites (c, XPL). Plagioclase forms subhedral crystals with oscillatory zoning and exhibits crystal sizes smaller than those of the K-feldspar and quartz. K-feldspar is perthitic and presents uniform alteration with formation of clay minerals.

Reactivity potential attributed to:
Abundant microcrystalline quartz (myrmekites).

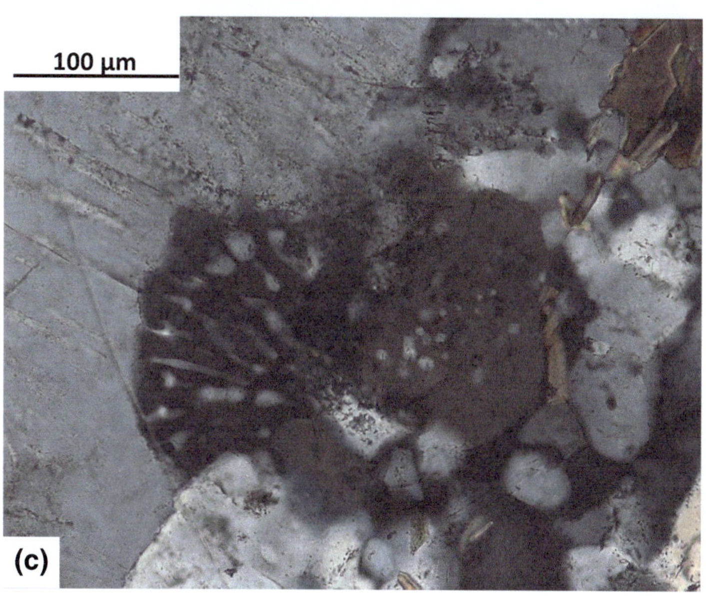

Sample of an aggregate that proved to be reactive in a concrete dam (Spain).

Granite (PT)

Slightly to moderately weathered two-mica granite composed of quartz, K-feldspar (perthitic microcline), plagioclase, muscovite, and biotite. It contains iron oxides, titanite, rutile, apatite, zircon, andalusite, chlorite, epidote, sillimanite (fibrolite) as accessories. The rock has hypidiomorphic inequigranular texture. Plagioclase forms large crystals, mainly zoned and weathered in their nucleus. Numerous cracks filled with iron oxides cross the quartz and the feldspar laths (a, PPL). Quartz occurs in well-developed crystals with slight undulatory extinction (<15°), showing frequent microcracking. Sericite crystals develop in the interstitial spaces, forming clusters of fibrous crystals (b, XPL). The larger crystals of quartz show sutured grain boundaries and microcrystals, localy with interlobate limits (c, XPL).

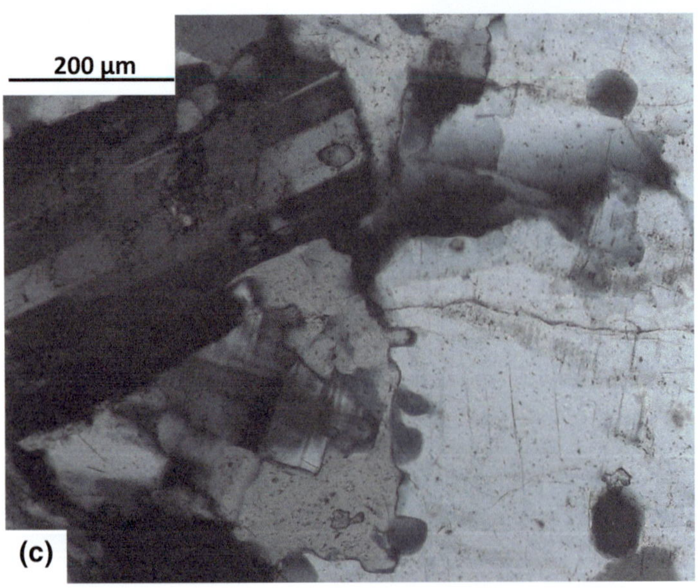

Reactivity potential attributed to:
Crystals of quartz have indented grain contacts. Abundant fractures cross coarse quartz and/or feldspar crystals.

Aggregate particle in a sample of concrete with ASR from a concrete dam (Portugal).

Granite (PT)

Deformed two-mica granite composed of quartz, K-feldspar, plagioclase, muscovite, minor biotite, chlorite and opaque minerals. The rock shows hypidiomorphic inequigranular texture with preferred orientation of the minerals, especially the mica flakes (a, XPL). Quartz occurs as stretched and elongated microcrystals in domains aligned parallel to the oriented mica flakes. Larger quartz grains show undulatory extinction (<15°) with sutured grain boundaries and microcrystals. Myrmekites are common (b, XPL). K-feldspar (orthoclase) forms subhedral grains and also occurs in the groundmass. Elongated clusters of biotite and platy muscovite crystals are observed, occasionally with bent cleavage planes (c, XPL).

Reactivity potential attributed to:
Domains of elongate microcrystalline quartz with sutured boundaries. Abundant myrmekites.

Aggregate particle in a sample of concrete with ASR from a dam 40 years after construction (Portugal).

Granite (UK)

Two-mica granite composed of quartz, K-feldspar, minor plagioclase, muscovite, biotite (with trace amounts of chlorite and tourmaline as accessories). The rock exhibits hypidiomorphic inequigranular texture. The feldspar crystals are cloudy from partial alteration to kaolinite. Small sized crystals of zoned tourmaline (strongly green to brownish pleochroic) occur very often in the interstitial spaces between quartz and feldspar (a, PPL). Quartz occurs in developed anhedral crystals with uniform extinction (b, XPL). Plagioclase forms suhedral to anhedral laths and constitutes the most strongly altered mineral in the rock (c, XPL).

Reactivity potential attributed to:
With this type of rock and texture, the reactive component is hard to identify. The crystals of quartz exhibit lightly curved boundaries.

Aggregate particle in a sample of concrete from a bridge (United Kingdom).

Granite (UK)

Granite composed of quartz, K-feldspar, plagioclase, muscovite and biotite (with trace amounts of chlorite, rutile and zircon as accessories).The rock exhibits hypidiomorphic inequigranular texture, in which the main components occur as anhedral crystals with straight boundaries (a, XPL). Smaller grains of the main constituent minerals are also present at the boundaries between coarser individuals. Quartz is abundant: the crystals are clear and show straight to slightly curved boundaries. Concave-convex limits are observed locally (b, XPL). Plagioclase is partly altered to sericite, K-feldspar (orthoclase) to kaolinite, biotite flakes to chlorite. Muscovite flakes are secondary, coarser than biotite, growing between quartz and feldspar (c, XPL).

Reactivity potential attributed to:
With this type of rock and texture, the reactive component is hard to identify. The crystals of quartz exhibit slightly curved boundaries.

(a)

(b)

(c)

Aggregate particle in a sample of concrete from a bridge (United Kingdom).

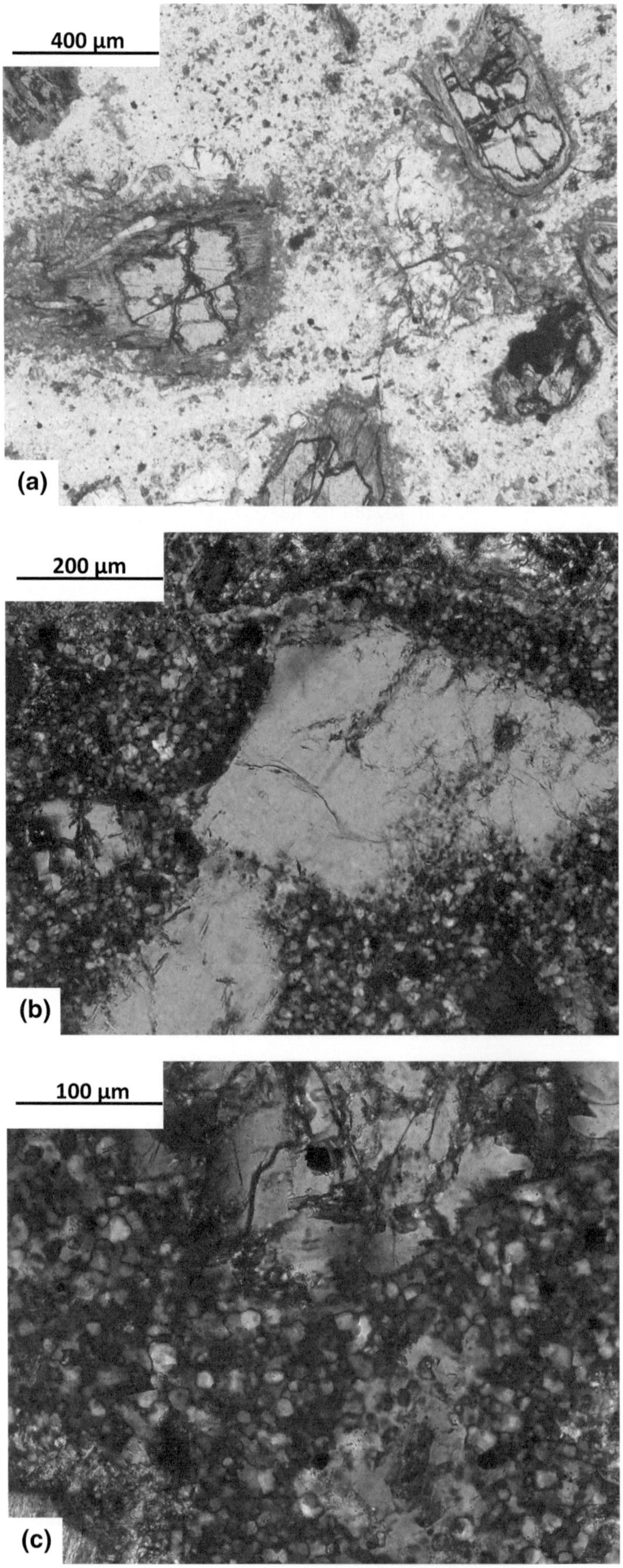

Rhyodacite (AU)

Porphyritic rhyodacite composed of plagioclase, K-feldspar, and quartz in roughly equal proportions, augite, hornblende, and chlorite (vermiculite), with trace amounts of apatite, zircon, and opaque minerals. Augite occurs in large phenocrysts, which margins and fractures are partially altered to hornblende, and further to vermiculite (a, PPL). Opaque minerals are closely associated with the original augite. Euhedral phenocrysts of zoned plagioclase and K-feldspar appear cloudy due to alteration mainly along cracks (b, XPL). The matrix is very fine-grained and contains microcrystalline quartz, K-feldspar and plagioclase, very difficult to distinguish from one another due to absence of twinning (c, XPL).

Reactivity potential attributed to:
Microcrystalline to cryptocrystalline quartz in the devitrified groundmass, cryptocrystalline quartz as a by-product of alteration.

Aggregate particle in a concrete sample (Australia).

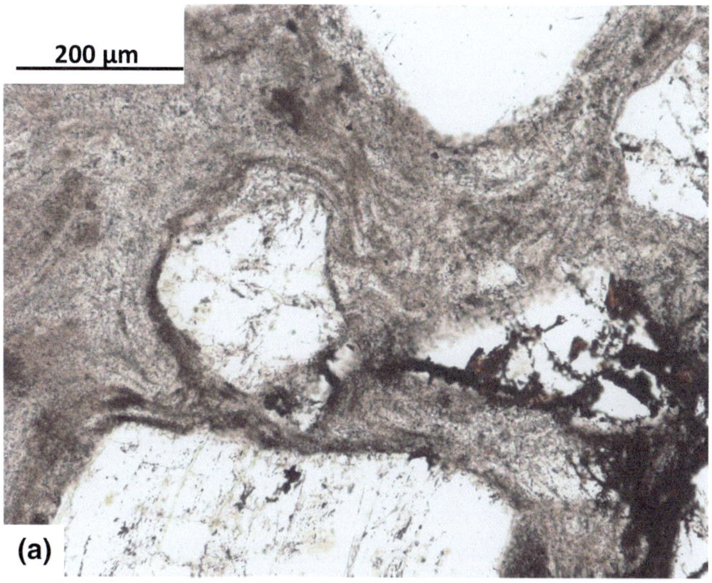

(a)

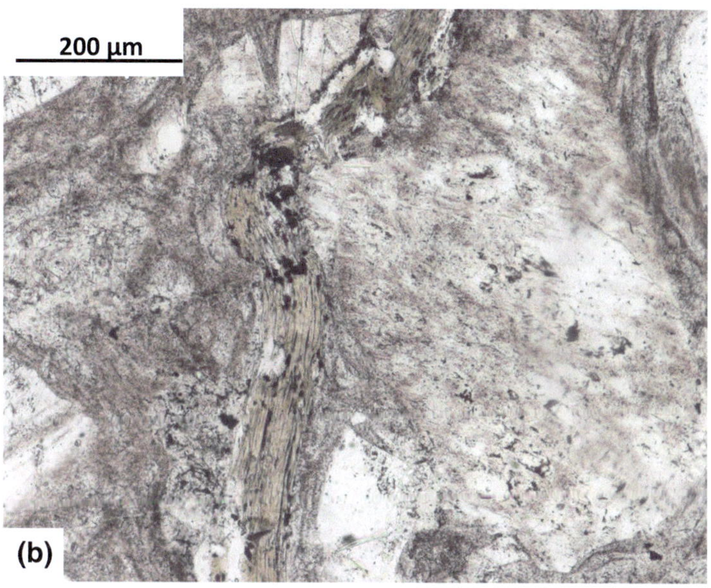

(b)

(c)

Rhyolite (AR)

Rhyolite showing porphyritic fluidal texture and composed of quartz, alkaline feldspar (sanidine), minor plagioclase (albite-oligoclase), and trace amounts of biotite, hornblende, titanite, apatite, zircon and opaque minerals. This altered felsic volcanic rock contains abundant phenocrysts dispersed in holohyaline (glassy) to microlitic matrix (a, PPL). Quartz occurs as phenocrysts, usually euhedral, and K-feldspar, also as phenocrysts, shows alteration and albitization. Plagioclase is less frequent and forms phenocrysts typically altered. Some phenocrysts of feldspar contain melt/volcanic glass inclusions. Biotite occurs in elongated lamellae, some along phenocrysts edges (b, PPL). The matrix shows rare microcrysts, but is predominantly holohyaline. It has a high content of volcanic glass (isotropic) and microcrystalline alkaline feldspar (c, XPL).

Reactivity potential attributed to:
Glassy matrix with some devitrification resulting in cryptocrystalline felsitic aggregations composed of quartz or tridymite and chalcedony (according to literature).

Sample of an aggregate that proved to be reactive in concrete pavements (Argentina).

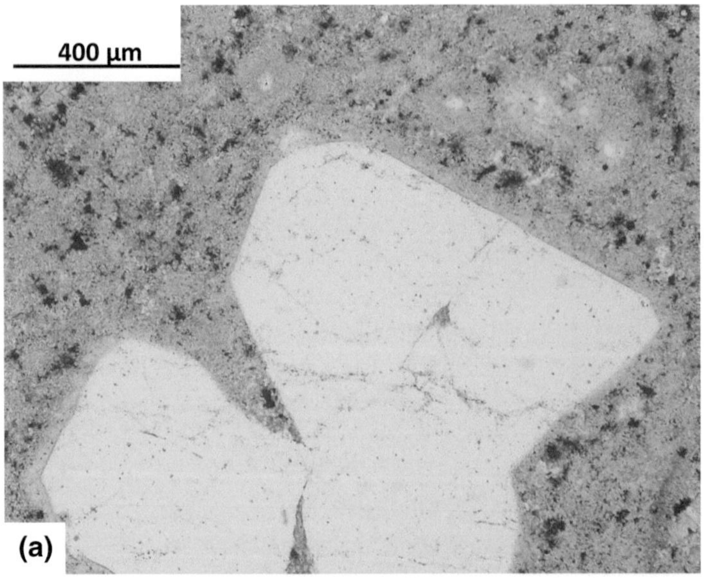

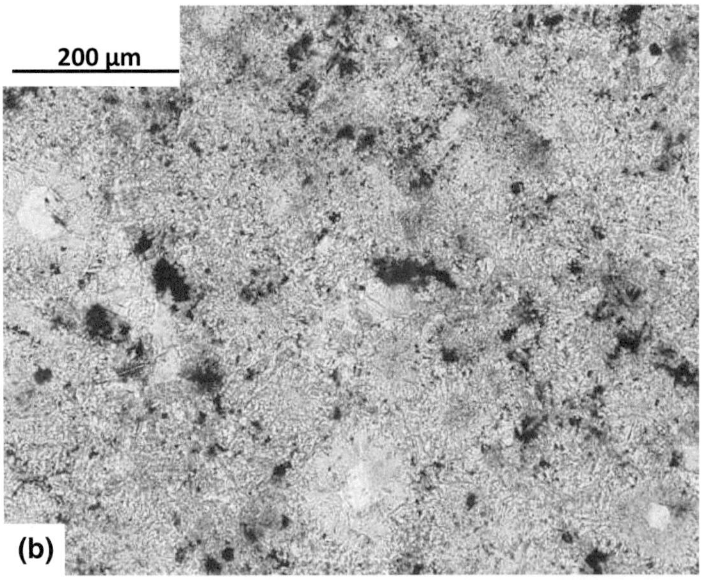

Rhyolite (DE)

Rhyolite showing a porphyritic texture and composed of quartz, K-feldspar (sanidine), minor plagioclase (albite-oligoclase), and trace amounts of biotite, hornblende (some crystals deeply chloritised) and opaque minerals. The phenocrysts of quartz and plagioclase are distributed throughout a hypocrystalline matrix. Quartz phenocrysts (a, PPL) are larger than plagioclase, and are very often euhedral with corrosion gulfs. Phenocrysts of amphibole and biotite are much rarer, and are partially chloritised. Opaque minerals occur finely dispersed throughout the rock.
The cryptocrystalline groundmass is composed of devitrified glass and contains quartz and feldspar, in some places showing spherulitic texture (radiating fibres due to the recrystallization of glass) (b, PPL). The groundmass also contains volcanic glass (c, XPL).

Reactivity potential attributed to:
Glassy matrix with some devitrification resulting in cryptocrystalline felsitic aggregations composed of quartz (or tridymite) and chalcedony.

Aggregate particle in a sample of concrete from a hydraulic structure (Germany).

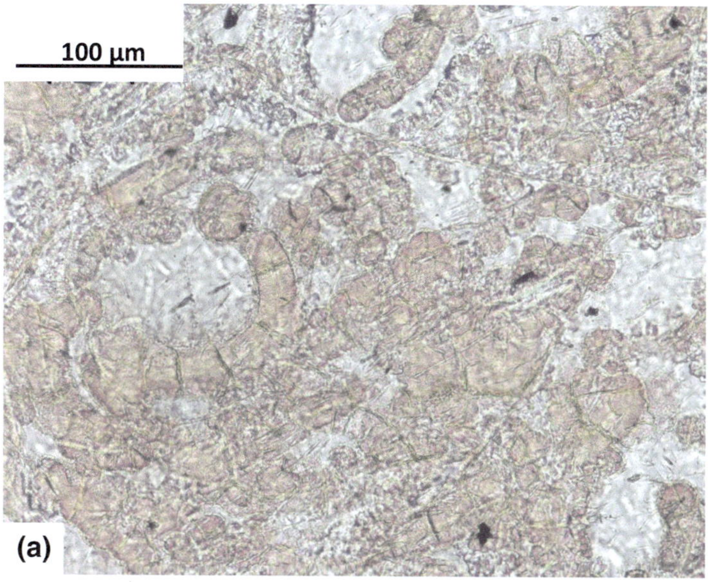

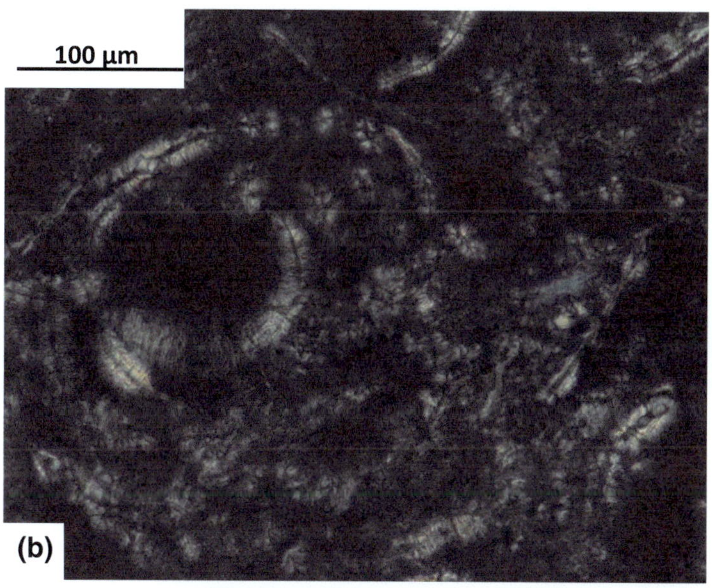

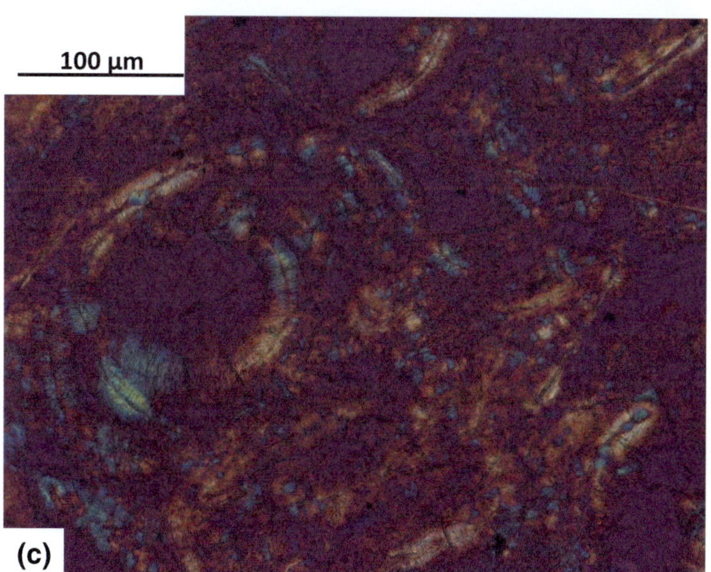

Rhyolite (IS)

Felsic volcanic rock with vitreous texture, and composed of quartz, feldspar and volcanic rhyolitic glass. This rhyolite shows a pseudo-perlitic texture with concentric cracks (a, PPL). There are radiating fibres of feldspar and minute interstitial tridymite or cristobalite with vermicular (axiolites) and amorphous groundmass. Cristobalite, tridymite and/or amorphous silica may be present in the glassy groundmass, as well as clay minerals from alteration. Due to the abundance of isotropic glassy material (b, XPL) the gypsum plate was used in order to emphasise the fibrous texture of the axiolites. In this image (c, GPXPL), the dark-red colour reveals the presence of amorphous silica.

Reactivity potential attributed to:
Rhyolitic volcanic glass and cryptocrystalline silica.

Aggregate particle in a concrete prism sample (Iceland).

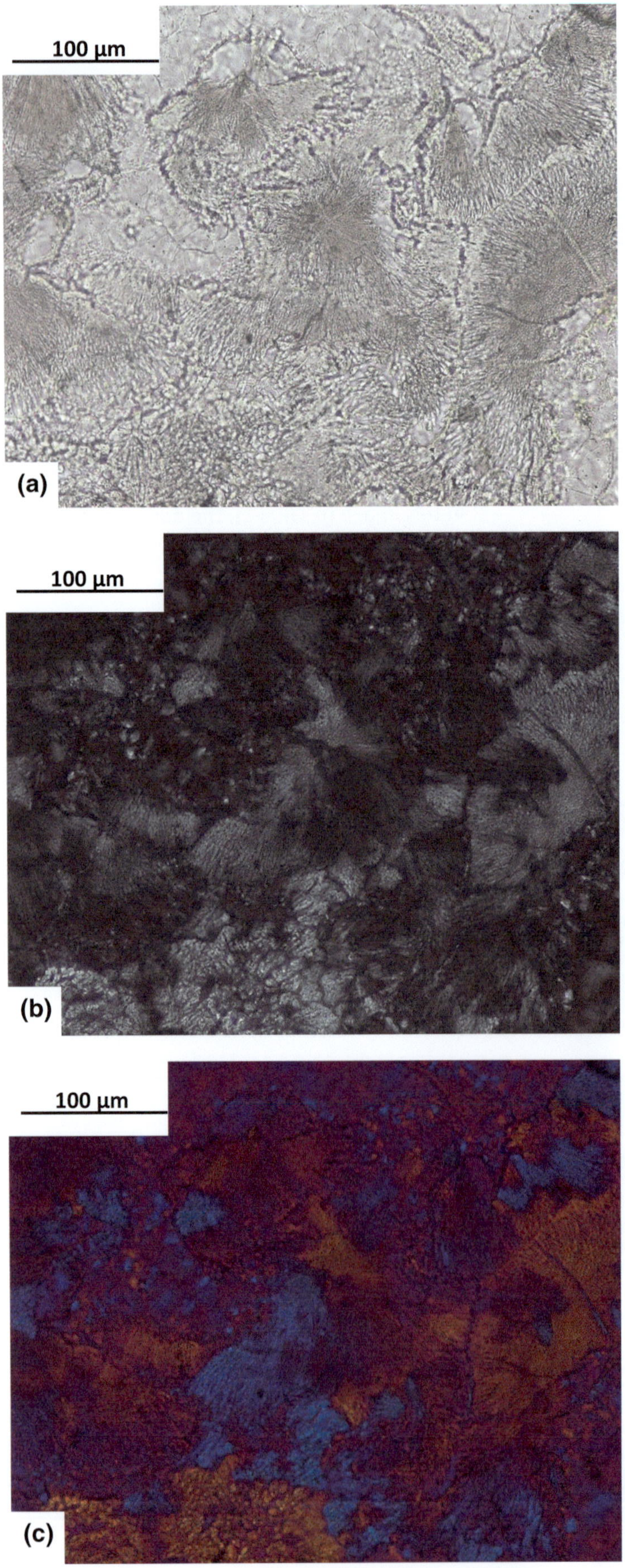

Rhyolite (IS)

Dense and unaltered rhyolite showing vitreous (perlitic) and vitro-clastic textures. This cryptocrystalline felsic rock is composed of quartz, feldspar and volcanic rhyolitic glass (a, PPL). Aggregations composed mainly of quartz or tridymite and chalcedony form the rock. Under cross polars, cryptocrystalline quartz and small laths of plagioclase are distinguished. The perlitic texture (cemented spherules) includes radiating fibres, with vermicules (axiolites) (b, XPL). The abundant intergrowths of felsic minerals result from the recrystallization (devitrification) of the groundmass. The intergrowths show radiating texture of very slim crystals (c, GPXPL).

Reactivity potential attributed to:
Rhyolitic volcanic glass and cryptocrystalline silica.

Aggregate particle in a sample of concrete prism sample (Iceland).

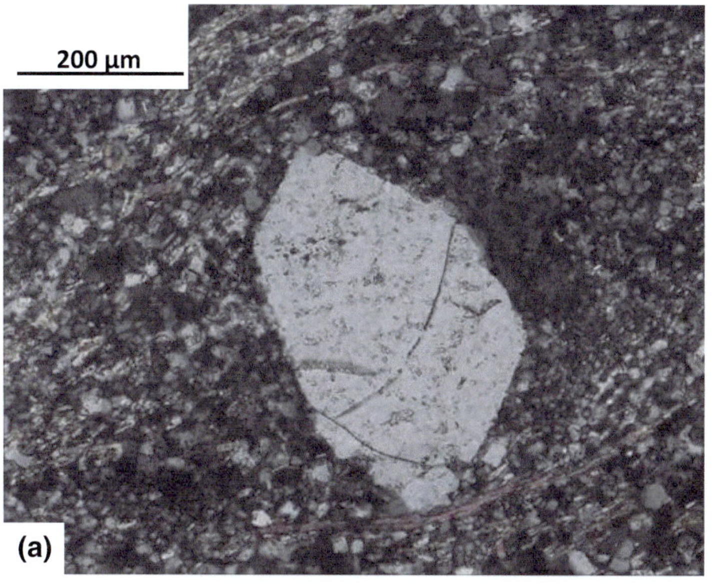

Rhyolite (NO)

Dense rhyolite composed of quartz and feldspar surrounded by a fine-grained groundmass of recrystallized volcanic (rhyolitic) glass. Euhedral megacrysts of quartz and feldspar are embedded in a groundmass (a, XPL) composed of crypto-microcrystalline feldspar, quartz and sericite. Flow direction is marked by the presence of parallel oriented micas and quartz bands, defining a preferred orientation (b, XPL). By contrast, quartz phenocrysts are not aligned with the groundmass (c, XPL).

Reactivity potential attributed to:
Microcrystalline to cryptocrystalline silica.

Aggregate particle in a sample of concrete prism. This aggregate showed to be reactive also in hydro power plants and dams (Norway).

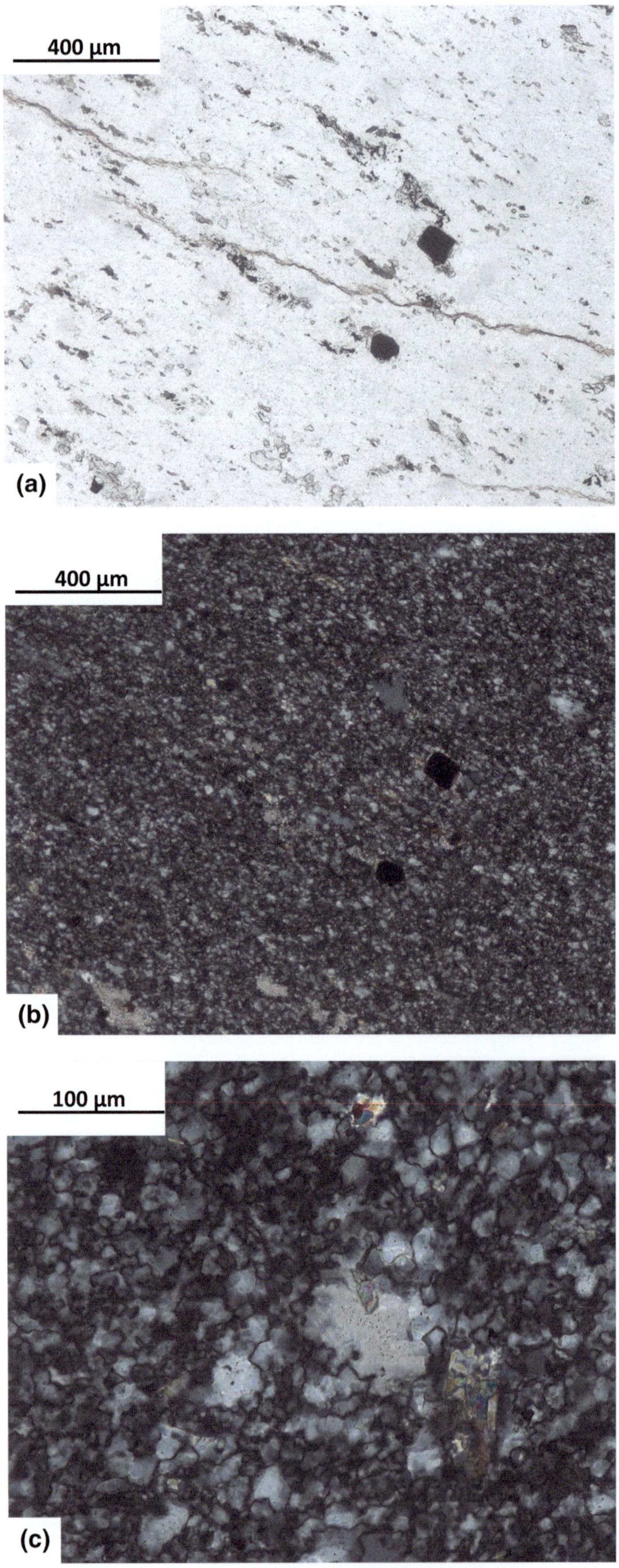

Rhyolite (SE)

Rhyolite with microlitic flow texture and preferred orientation of the minerals (a, PPL). It contains phenocrysts of quartz, K-feldspar and plagioclase (difficult to distinguish due to fine-grained nature of the rock), minor biotite (secondary), and traces of opaque minerals, calcite (secondary), chlorite and titanite. Secondary calcite appears slightly aligned in the groundmass (b, XPL). Phenocrysts are embedded in a recrystallized hypocrystalline matrix composed mainly of quartz and feldspar. The larger quartz grains in the groundmass show irregular boundaries (c, XPL).

Reactivity potential attributed to:
Microcrystalline to cryptocrystalline quartz, sometimes fibrous (chalcedony), in the groundmass.

Aggregate particle in a concrete prism sample (Sweden).

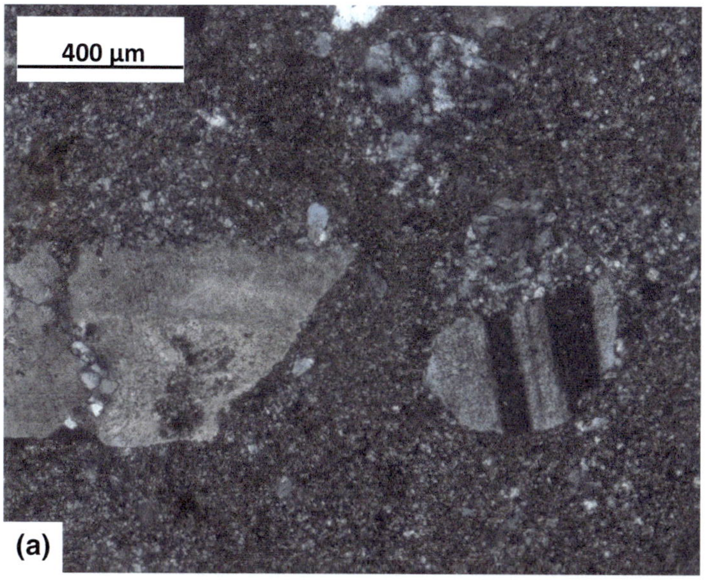

Rhyolite (SG)

Rhyolite showing a porphyritic texture and composed of quartz, K-feldspar, minor plagioclase, and trace amounts of biotite (of secondary origin) and opaque minerals. Euhedral phenocrysts of quartz and feldspar are embedded in a hypocrystalline groundmass (a, XPL). The rock is altered as shown by brownish staining (b,PPL). This felsic volcanic rock shows abundant cryptocrystalline quartz in the groundmass. The phenocrysts may show corrosion gulfs (c, XPL).

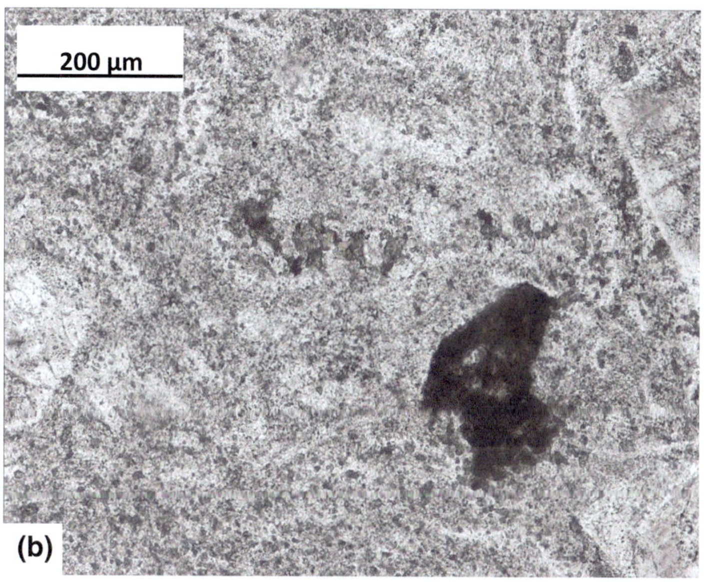

Reactivity potential attributed to:
Microcrystalline to cryptocrystalline quartz in the groundmass.

Sample of rock that proved to be reactive in laboratory tests (Singapore).

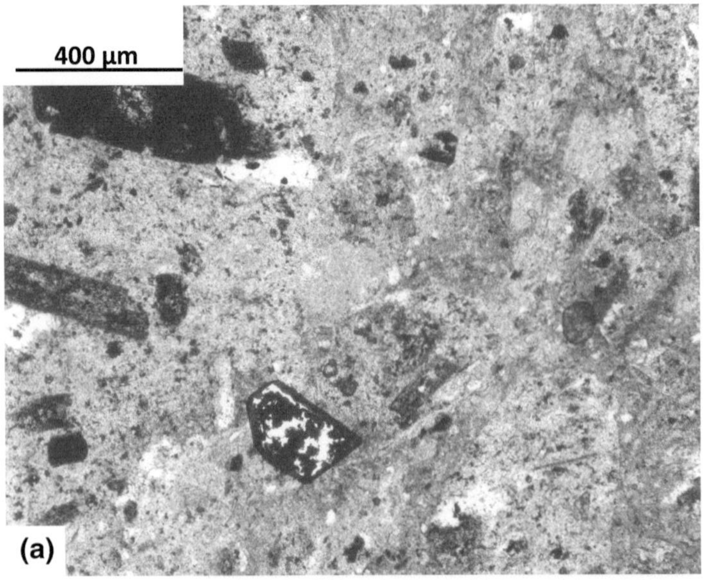

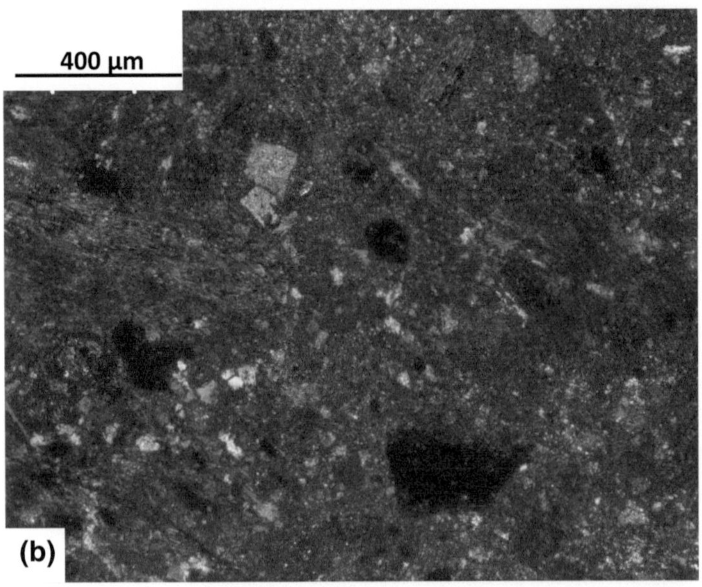

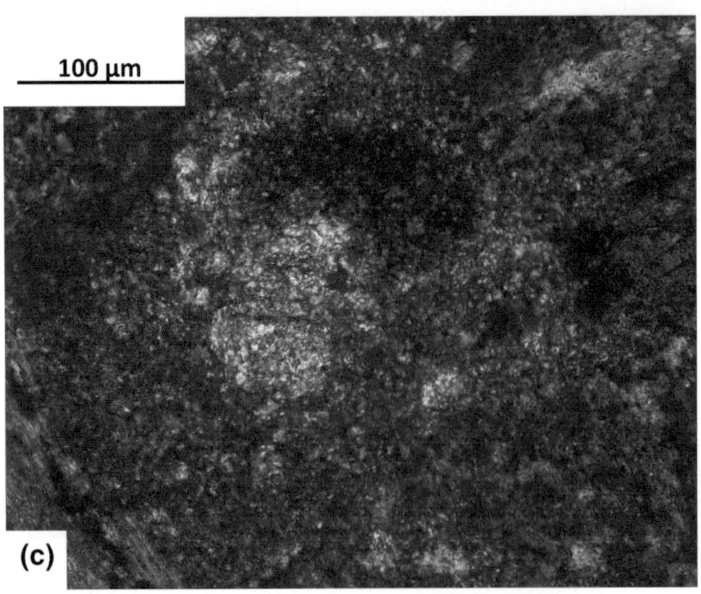

Rhyolite (USA)

Felsic volcanic rock showing microlitic flow texture and composed of quartz, K-feldspar (sanidine), minor plagioclase (albite-oligoclase), biotite (partially chloritised), hornblende (some crystals intensely chloritised) and trace amounts of opaque minerals. This rhyolite contains phenocrysts dispersed in the hypocrystalline to glassy matrix (a, PPL). Quartz and plagioclase form phenocrysts and also compose most of the recrystallised groundmass. The rock is altered: chlorite is observed to replace biotite and amphibole phenocrysts (b, XPL). The isotropic character of the groundmass reveals the presence of volcanic glass, although there is also formation of cryptocrystalline quartz, feldspar and sericite (c, XPL).

Reactivity potential attributed to:
Glassy matrix. The matrix is composed of volcanic glass partially recrystallized to cryptocrystalline quartz and feldspar.

Aggregate particle in a concrete prism sample. This aggregate showed also to be reactive in a highway structure (USA).

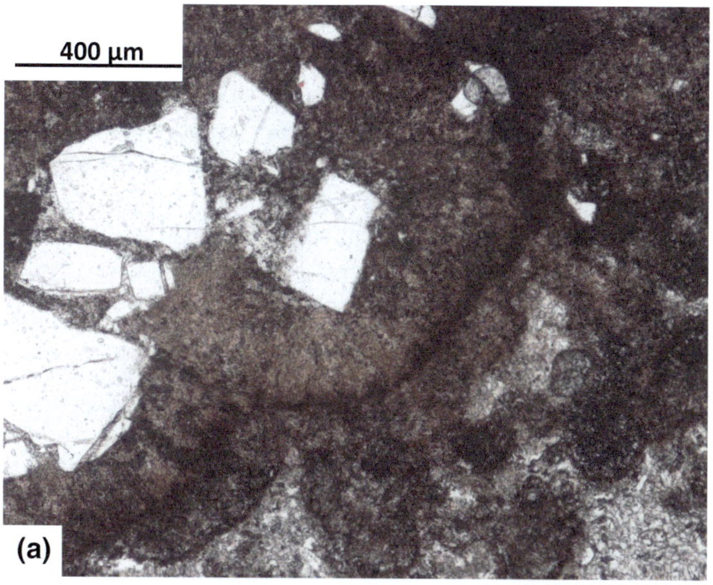

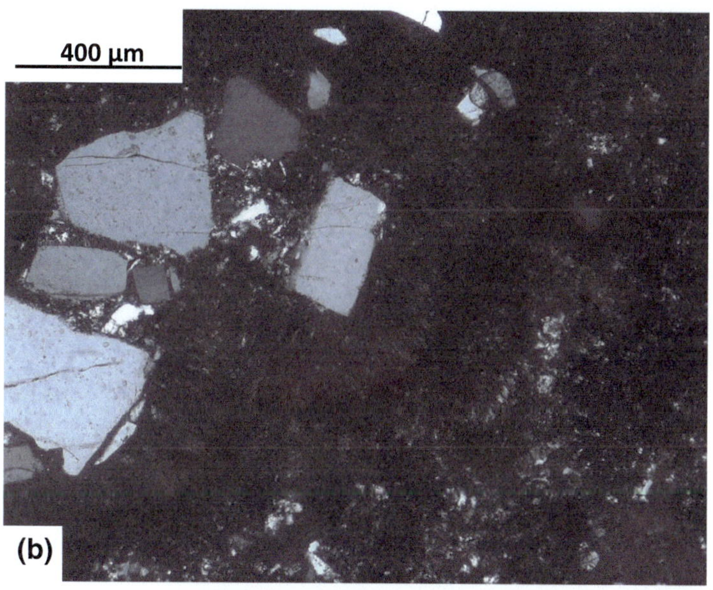

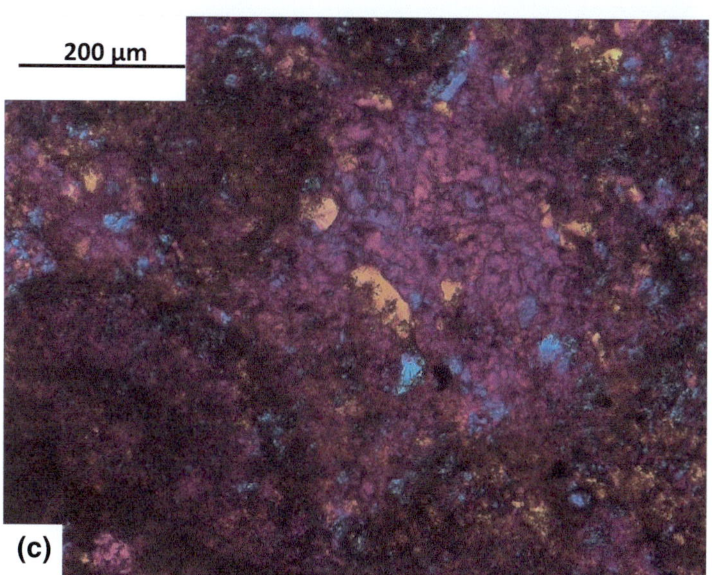

Rhyolite (USA)

Felsic volcanic rock showing vitro-clastic texture composed of phenocrysts of quartz, K-feldspar (orthoclase and microcline), plagioclase and opaque minerals. Quartz occurs in the rhyolite groundmass, besides feldspars and volcanic glass. The phenocrysts are scarce and unevenly distributed. Spherules of volcanic glass (dark brown), with rims of iron hydroxide, are abundant (a, PPL). The groundmass is cryptocrystalline and locally spherulitic, interstitial space being filled with microcrystalline to cryptocrystalline quartz and feldspar. Cavities contain concentric or radiating fibro-crystalline quartz and feldspar (b, XPL). Due to the isotropic character of the glass the gypsum plate was used, thus revealing the dark-red colour characteristic of amorphous material (c, GPXPL).

Reactivity potential attributed to:
Volcanic glass (composition should be accessed by EDS or WDS). Amorphous silica and cryptocrystalline quartz in the groundmass.

Aggregate particle in a concrete prism sample (USA).

Tuff (NZ)

Fine-grained, volcaniclastic rock composed of welded lithic fragments of acid volcanic rock (rhyolite), amphibole, quartz and K-feldspar (a, PPL). Coarse fragments are angular, consolidated and welded together in a dark brownish, isotropic glassy matrix from very-fine grained ash material, stained with 'limonite' (b, PPL). Non-silicic fragments appear cloudy from advanced alteration (c, XPL), forming various products including cryptocrystalline quartz and clay minerals.

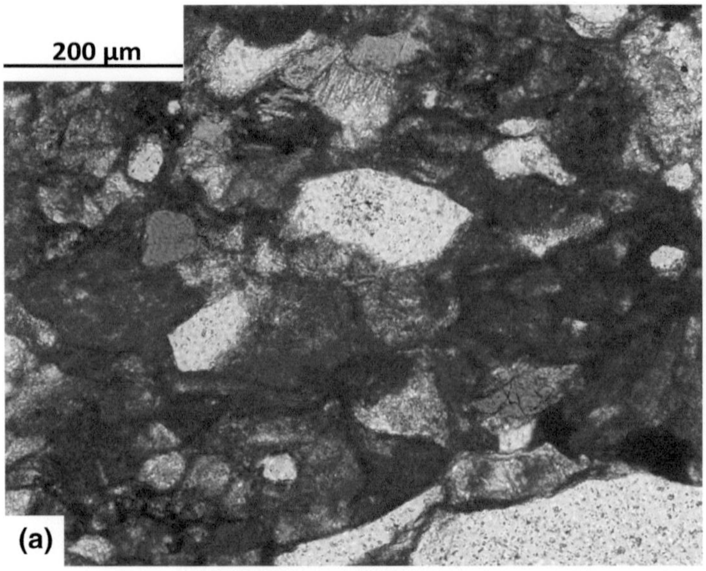

Reactivity potential attributed to:
Cryptocrystalline quartz in the groundmass, and as by-product from mineral alteration.

Aggregate particle in concrete from a structure (New Zealand).

Tuff (NZ)

Fine-grained, volcaniclastic rock composed of quartz, K-feldspar, minor welded lithic fragments (rhyolite, calc-silicate rock), and traces of opaque minerals (a, PPL). Coarse angular fragments are welded by a dark brownish, isotropic glassy matrix from very-fine grained ash material, stained with 'limonite' (b, PPL). Non-silicic fragments appear cloudy from advanced alteration, forming various products including cryptocrystalline quartz and clay minerals (c, XPL).

Reactivity potential attributed to:
Cryptocrystalline quartz in the groundmass and lithic fragments, and as by-product from mineral alteration.

Aggregate particle in concrete from a structure (New Zealand).

Sedimentary Rocks

3

Isabel Fernandes, Maarten A.T.M. Broekmans, Maria dos Anjos Ribeiro, and Ian Sims

Sedimentary rocks form from deposition, compaction and consolidation of fragments, minerals and ionic solutions of older rocks near or at the Earth's surface in atmosphere or underwater, normally deposited in horizontal layers. The composing materials may have different sources: (1) lithic or mineral detritus from natural breakdown of pre-existing rocks (igneous, sedimentary or metamorphic) through weathering, erosion, transport, and deposition (clastic sediments), (2) chemical sediments formed by precipitation from solution (non-clastic), or (3) sediments of organic origin containing plant or animal remains and/or secretions. Compaction and consolidation occur in the upper crust, but by definition exclude formation of metamorphic minerals.

Sedimentary rocks are used as aggregates for concrete in many countries all over the world both as coarse and fine aggregates. Several papers have been published on alkali-silica reaction of sedimentary rocks in concrete causing damage (e.g. Oberholster 1981, 2001; Cole and Lancucki 1983; Sims 1992; Shayan and Ferguson 1996; Freitag et al. 2000; Guédon-Dubied et al. 2000; Heijnen et al. 1996; Marfil et al. 2012; Sanchez et al. 2012). Among the best known examples of alkali-reactive sedimentary lithologies are the crushed greywacke from the UK (Sims 1992), the siliceous limestone from Spratt's (Fournier and Rogers 2008; Rogers and MacDonalds 2012), and the chert and opaline lithologies in aggregates from the North of Europe (Sprung and Sylla 1998; Franke et al. 2004; Lorenzi et al. 2006).

The most common consolidated clastic sedimentary rocks used as aggregate for concrete are greywacke, various types of sandstone and, in a much lower proportion, some otherwise suitable types of mudstone (siltstone and claystone). These rocks are predominantly composed of siliceous detritus ('clasts') consolidated by subsequent compaction and cementation during diagenesis. Compaction originates pressure-solution, with the solution and transport of reaction products created by differential stress. This process is observed in well sorted sandstones with the indentation of the clasts in the direction where the main stress is acting and deposition of silica in the direction of the lower stress. The microscopic features characteristic of pressure/solution include the growth of the grains in which the original boundaries are outlined by oxides or impurities (e.g. Chester et al. 2004; Vernon 2011). Where pressure-solution is more intense, the contacts between grains become sutured.

The space between the clasts may be filled by smaller particles (detrital mud) or by chemical deposition of compounds. The matrix is the detrital fraction composed of smaller particles (grains <30 μm) which occupy the intergranular spaces and may differ greatly from the sedimentary clasts (Poole 1992). It is mainly composed of clay minerals and/or microcrystalline to cryptocrystalline quartz and may be recrystallized to secondary micas or chlorite when the rock has been subject to low-grade metamorphic conditions.

Chemical compounds may precipitate as new minerals (rock cement) in spaces between the fragments, forming a solid sedimentary rock by cementing the clasts together. The most common rock cement minerals are silica, carbonates, hematite, limonite, chlorite and clay minerals. Interstitial/intergranular pore space between the detrital grains is often observed open and connected, facilitating access of concrete pore fluid to the rock's interior.

This document is not intended to discuss further the ideas about sedimentary processes and rocks for which the reader

I. Fernandes (✉)
ICT, Institute of Earth Sciences and Department of Geology, Faculty of Sciences, University of Lisbon, Lisbon, Portugal
e-mail: mifernandes@fc.ul.pt

M.A.T.M. Broekmans
Department of Mineral Resources, Geological Survey of Norway - NGU, Trondheim, Norway
e-mail: maarten.broekmans@ngu.no

M.A. Ribeiro
ICT, Institute of Earth Sciences and DGAOT, Faculty of Sciences, University of Porto, Porto, Portugal
e-mail: maribeir@fc.up.pt

I. Sims
RSK Environment Ltd, Hemel Hempstead, UK
e-mail: ian@simsdoc.com; isims@rsk.co.uk

I. Fernandes et al. (eds.), *Petrographic Atlas: Characterisation of Aggregates Regarding Potential Reactivity to Alkalis*, DOI 10.1007/978-94-017-7383-6_3

can get information from specialised literature. In this atlas the classification presented by Dott (1964) and Pettijohn (1983) has been followed in which the division for sandstones is based on the content of matrix as: sandstones (<15 %), wackes (15–75 %) and mudstones (>75 %), the nature of the clasts and their relative content (quartz, feldspar or rock fragments) and the grain size.

The samples received from many different areas allow the identification of the main reactive consolidated sedimentary rocks. The most common potentially reactive detrital rocks presented here are mainly composed of clasts of quartz and an argillaceous matrix, in some cases recrystallized to biotite, sericite or chlorite. Clay may be allogenic (clastic or detrital) or authigenic (rock cement or replacement of feldspars). Microcrystalline quartz exists in the matrix and also at the grain boundaries. It can also form micro-vugs, when quartz grains nucleate on the clasts and develop perpendicular to the clast surfaces, and may nucleate in the same crystallographic orientation as clasts of the same mineral forming pressure overgrowths (secondary enlargements), resulting in a mosaic which tends to have polygonal shapes.

Some cases of siliceous limestones are also included. The objective of the present work is not concerned with the occurrence of alkali-carbonate reactions, the nature of which is still a matter of dispute (e.g. Katayama 2004). However, recently 'cryptic quartz' (i.e. undetectable in thin section petrography) present in the carbonate mud has been identified as responsible for the deleterious reaction of certain limestones (Grattan-Bellew et al. 2010; Katayama 2010) (e.g. Kingston limestone). Thus, at least in some cases the expansion associated with carbonate rocks has been shown to be a special case of ASR. Although the identification of the reactive silica in such cases usually requires liberation through selective dissolution in excess acid and XRD analysis of the insoluble residue, photomicrographs are included of some of siliceous limestones known to cause deterioration both in field structures and in laboratory tests. In some cases also SEM/EDS was applied in order to observe the texture and to identify the occurrence of microcrystalline silica.

Chert is a chemically precipitated sedimentary rock, essentially monomineralic and composed of microcrystalline quartz and chalcedony with minor amounts of impurities such as clay minerals, silt, carbonate, pyrite and organic matter and, in some cases, opal (Folk 1980). Chert can be composed of authigenic silica due to precipitation from the hydrous solution or may be secondary in the sediment. When it is primary, it forms from the organic remains and fragments originated from siliceous shells, siliceous spiculae, radiolaria and diatoms. Some cherts include iron oxides. Secondary cherts formed by replacement of carbonate rocks may develop as nodules or layers. Chalcedony may fill voids or replace the originally-formed silica.

In some areas of the world, such as north-western Europe, flint/chert occur in certain natural sand and gravel deposits that have been widely exploited for concrete aggregate. Flint/chert is composed of microcrystalline to cryptocrystalline quartz (Heaney and Post 1992; RILEM 2016, see Nixon and Sims 2016), and microporous varieties are considered particularly alkali-reactive, even at low modal contents (pessimum effect). In the absence of information to the contrary, all chert should be regarded as potentially deleteriously reactive with the alkalis in hydraulic cement paste (Alexander and Mindess 2005). Their reactivity can be attributed to the fine-grained nature of the quartz and areas of defective lattice, as well as to the variable presence of silica varieties such as chalcedony, opal (Winkler 1979; Passchier and Trouw 1996), and quite possibly moganite (Broekmans 2004), a little known silica polymorph apparently common in flint/chert worldwide (Heaney and Post 1992). Although the mineral composition is only slightly variable, cherts from a number of samples and countries are included in order to exemplify different textures and features of these rocks.

Chert/flint show, in some cases, a correlation between porosity and alkali-reactivity but in other cases it has been found that high porosity cherts behave as non-reactive in expansion tests (Barisone and Restivo 2000). Internal porosity seems to be a crucial factor, and can be identified using fluorescence-impregnation petrography and scanning electron microscopy.

The reactivity of sedimentary rocks, whether siliceous or calcareous, is often attributed to the presence of opaline silica, or more commonly various forms of microcrystalline or cryptocrystalline quartz in chert/flint and siliceous limestones, or as chalcedony and possibly moganite.

In the descriptions and in the photomicrographs some low-grade metamorphic features can be found. However, whenever the original texture of the rock maintains the sedimentary characteristics, the sample was included in this chapter. In some ambiguous cases (owing to not having information about the geologic context of the rock) the reference to metamorphic features was included in the description of the sample.

References

Alexander, M, Mindess, S (2005): Aggregates in concrete. Modern Concrete Technology 13. Taylor & Francis. pp 435.

Barisone, G, Restivo, G (2000): Alkali-silica reactivity of some Italian opal and flints tested using a modified mortar bar test. In: Bérubé, MA, Fournier, B, Durand, B (editors), Proceedings of the 11th International Conference on Alkali-Aggregate Reaction (ICAAR), Québec, Canada: 239–245.

Broekmans, MATM (2004): Structural properties of quartz and their potential role for ASR. Materials Characterization (53): 129–140.

Chester, JS, Lenz, SC, Chester, FM, Lang, RA (2004): Mechanisms of compaction of quartz sand at diagenetic conditions. Earth and Planetary Science Letters (220): 435–451.
Cole, WF, Lancucki, CJ (1983): Products formed in an aged concrete. The occurrence of okenite. Cement and Concrete Research (13): 611–618.
Dott, RH (1964): Wacke, graywacke and matrix; what approach to immature sandstone classification?. SEPM Journal of Sedimentary Research (34/3): 625–632.
Folk, RL (1980): Petrology of sedimentary rocks. 2nd edition. Hemphill Publishing Corporation: pp 182.
Fournier, B, Rogers, C (2008): Multi-laboratory study of accelerated mortar bar test and concrete prism expansion tests at 38° and 60°C. In: Broekmans, MATM, Wigum, BJ (editors), Proceedings of the 13th International Conference on Alkali-Aggregate Reaction in Concrete (ICAAR), Trondheim, Norway: 647–656.
Franke, L, Eickemeier, K, Bosold, D (2004): The alkali reactivity of opaline sandstone in aggregates, Materials Science, Zement-Kalk-Gips International (57): 76–83.
Freitag, SA, St John, DA, Goguel, R (2000): ASTM C1260 and the alkali reactivity of New Zealand greywackes. In: Bérubé, MA, Fournier, B, Durand, B (editors), Proceedings of the 11th International Conference on Alkali-Aggregate Reaction (ICAAR), Québec, Canada: 305–313.
Grattan-Bellew, PE, Mitchell, LD, Margeson, J, Min, D (2010): Is alkali–carbonate reaction just a variant of alkali–silica reaction ACR=ASR? Cement and Concrete Research (40/4): 556-562.
Guédon, S, Cadoret, G, Durieux, V, Martineau, F, Fasseu, P, Van Overbecke, V (2000): Study on Tournai limestone in Antoing Cimescaut Quarry. Petrological, chemical and alkali reactivity approach. In: Bérubé, MA, Fournier, B, Durand, B (editors), Proceedings of the 11th International Conference on Alkali-Aggregate Reaction (ICAAR), Québec, Canada: 335–344.
Heaney, PJ, Post, JE (1992): The widespread distribution of a novel silica polymorph in microcrystalline quartz varieties. Science (255): 441–443.
Heijnen, WNN, Larbi, JA, Siemes, AJM (1996): Alkali-silica reaction in The Netherlands. In: Shayan, A (editor), Proceedings of the 10th International Conference on Alkali-Aggregate Reaction (ICAAR), Melbourne, Australia: 109–116.
Katayama, T (2004): How to identify carbonate rock reactions in concrete. Materials Characterization (53/2–4), Special Issue (29): 85–104.
Katayama, T (2010): The so-called alkali-carbonate reaction (ACR) – its mineralogical and geochemical details, with special reference to ASR. Cement and Concrete Research (40/4): 643–675.
Lorenzi, G, Jensen, J, Wigum, B, Sibbick, R, Haugen, M, Guédon, S, Åkesson, U (2006): Petrographic atlas of the potentially alkali-reactive rocks in Europe. FP5 – Partner-project-GRD1-CT-2001-40103. Geological Survey of Belgium Professional Paper 2006/1-N°302, pp 63.
Marfil, S, Batic, O, Maiza, P (2012): Petrography of potentially alkali-reactive sandstone from Argentina. In: Drimalas, T, Ideker, JH, Fournier, B (editors), Proceedings 14th of the International Conference on Alkali-Aggregate Reactions in Concrete (ICAAR), Austin, USA, pp 10.
Nixon, PJ, Sims, I (2016): RILEM recommendations for the prevention of damage by alkali-aggregate reactions in new concrete structures, State-of-the-Art Report of the RILEM Technical Committee 219-ACS, RILEM State-of-the-Art Reports, Volume 17, Springer (Dordrecht, The Netherlands) pp RILEM (Paris).
Oberholster, RE (1981): Alkali-aggregate reaction in South Africa – a review. In: Oberholster, RE (editor), Proceedings of the 5th International Conference on Alkali-Aggregate Reactions in Concrete (ICAAR), Cape Town, South Africa. National Building Research Institute Pretoria. Paper S252/8: 1–11.
Oberholster, RE (2001): Alkali-silica reaction. In: Addis, BJ, Owens, G (editors), Fulton's Concrete Technology, 8th edition, Cement and Concrete Institute, Midrand, South Africa.
Passchier, CW, Trouw, RAJ (1996): Micro-tectonics. Springer Verlag, Heidelberg: pp 289.
Pettijohn, FJ (1983): Sedimentary rocks. 3rd edition, 2nd printing, Harper and Collins, New York: pp 628.
Poole, AB (1992): Introduction to alkali-aggregate reaction in concrete. In: Swamy, RN (editor): The alkali-silica reaction in concrete. Blackie & Son Ltd, London: pp 333.
RILEM (2016): RILEM Recommended Test Method: AAR-1.1 – Detection of Potential alkali-reactivity—part 1: petrographic examination method. In: Nixon, PJ, and Sims, I (editors): RILEM recommendations for the prevention of damage by alkali-aggregate reactions in new concrete structures. RILEM State-of-the-art Report (17): 35–60.
Rogers, C, MacDonald, C-A (2012): The geology, properties and field performance of alkali-aggregate reactive Spratt, Sudbury and Pittsburg aggregates distributed by the Ontario Ministry of Transportation. In: Drimalas, T, Ideker, JH, Fournier, B (editors), Proceedings 14th of the International Conference on Alkali-Aggregate Reactions in Concrete (ICAAR), Austin, USA, pp 10.
Sanchez, L, Salva, P, Fournier, B, Jolin, M, Pouliot, N, Hovington, A (2012): Evaluation of damage in the concrete elements of the Viaduct "Robert-Bourassa-Charest" after nearly 50 years in service. In: Drimalas, T, Ideker JH, Fournier, B (editors), Proceedings of the 14th International Conference on Alkali Aggregate Reactivity in Concrete (ICAAR), Austin, USA, pp 10.
Shayan, A, Ferguson, JA (1996): Reactive quartz gravel from Eastern Victoria. In: Shayan, A (editor), Proceedings of the 10th International Conference on Alkali-Aggregate Reaction (ICAAR), Melbourne, Australia: 703–710.
Sims, I (1992): Alkali-silica reaction - UK experience. In: Swamy, RN (editor): The alkali-silica reaction in concrete. Blackie & Son Ltd, London: pp 333.
Sprung, S, Sylla, H-M (1998): Course of the alkali-silica reaction in concrete with different types of aggregate, ZKG International (51): 334–346.
Vernon, RH (2011): A practical guide to rock microstructure. Cambridge University Press. pp 594.
Winkler, HGF (1979): Petrogenesis of metamorphic rocks. 5th Edition. Springer Verlag, New York: pp 348.

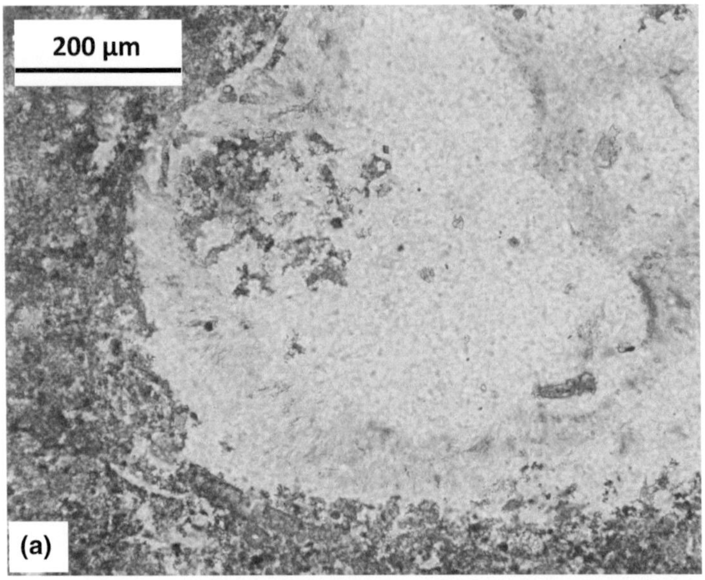

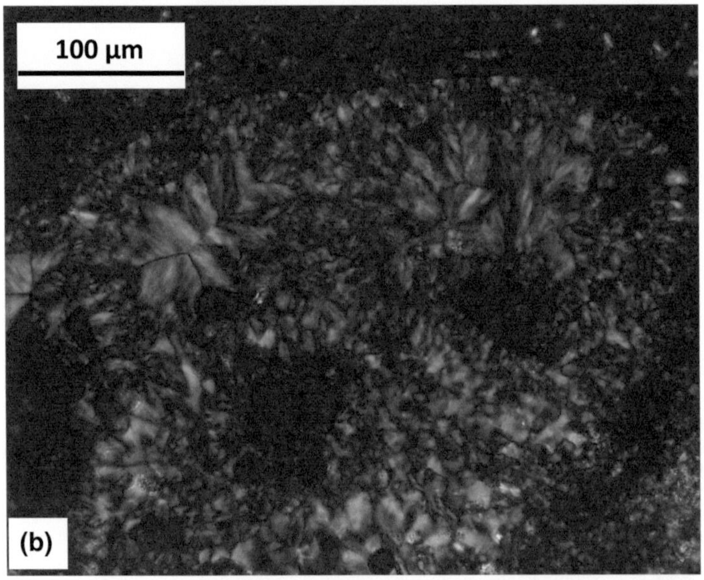

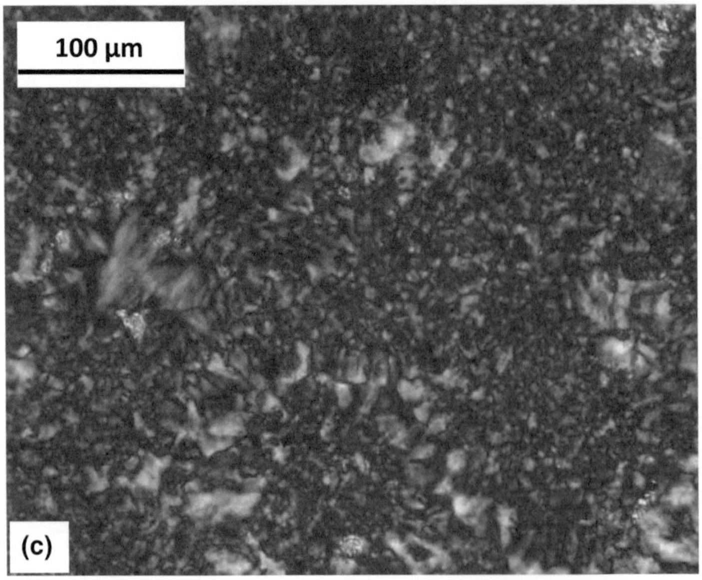

Chert (DK)

Opaline chert composed of cryptocrystalline quartz, opaline silica, chalcedony and trace amounts of calcite (a, PPL). Isotropic opaline silica is abundant in the matrix (colourless in PPL, isotropic in XPL). Interstitial limonite and/or clay minerals may be lining the internal microfabric of the chert. Various types of foraminifera and sponge spiculae can be recognized, among many fragmentary remains. The interior of fossils is often infilled by feathery chalcedony forming fans (b, XPL). Locally, traces of interstitial calcite may be present (c, XPL).

Reactivity potential attributed to:
Cryptocrystalline quartz, opaline silica, and chalcedony. This rock may display pessimum behaviour (according to literature).

Sample of rock classified as potentially reactive in laboratory tests and also in all types of structures such as bridges, swimming pool, car parks, balconies and marine structures (Denmark).

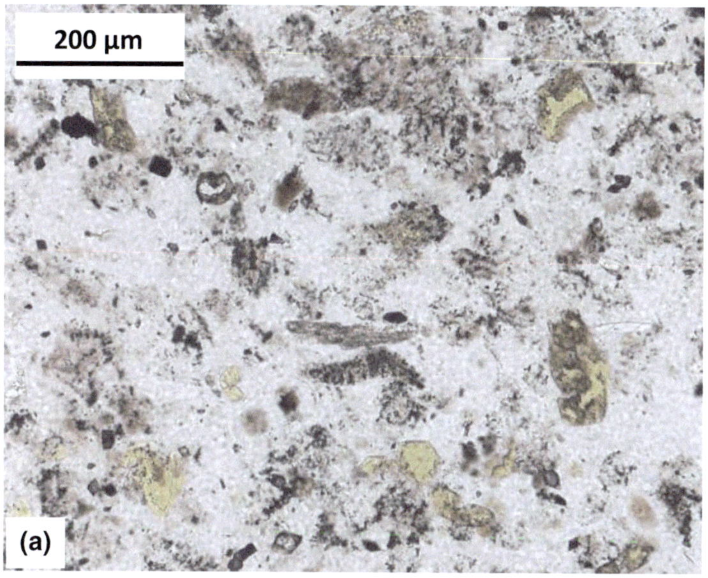

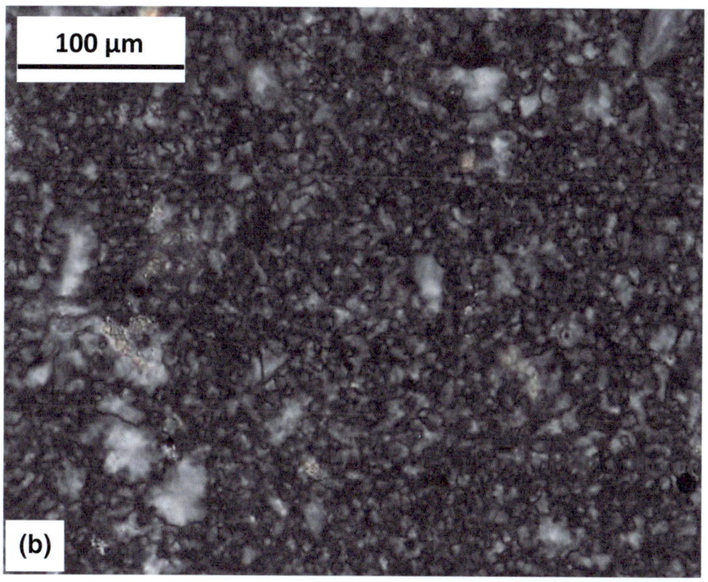

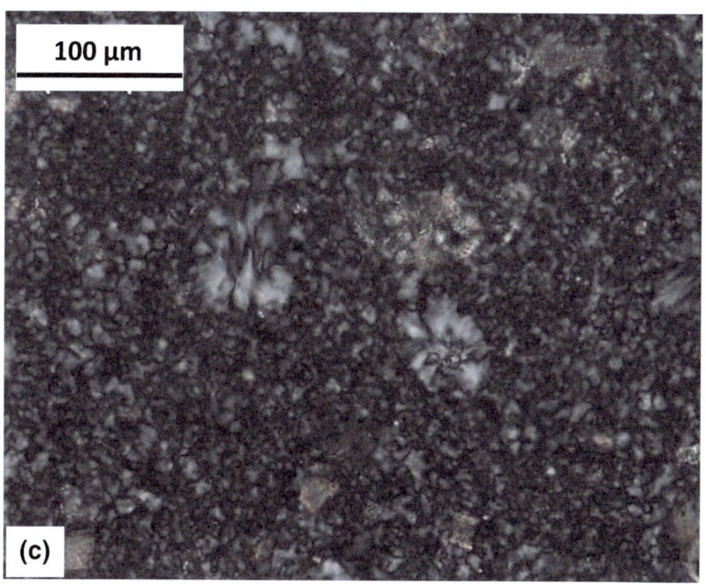

Chert (DK)

Porous chert composed of cryptocrystalline quartz, opaline silica, chalcedony, with finely interspersed organic matter, iron (oxy-) hydroxides (limonite) and carbonate. The chert contains various types of foraminifera, small rounded pellets (peloids), and fossil debris (bioclasts). The internal porosity of the chert is revealed by the yellow epoxy filling voids (a, PPL). Most fossil relics are filled with feathery chalcedony (b, XPL). The surrounding matrix is composed mainly of cryptocrystalline quartz. Locally, carbonate may occur as interstitial blobs (probably calcite), or as small lozenges characteristic of dolomite (c, XPL).

Note: this section was impregnated with a yellow-dyed epoxy for fluorescence.

Reactivity potential attributed to:
Cryptocrystalline quartz, chalcedony. This aggregate as sand may show pessimum behaviour (according to literature).

Constituent classified as potentially reactive in laboratory tests and in field structures and edifices (Denmark).

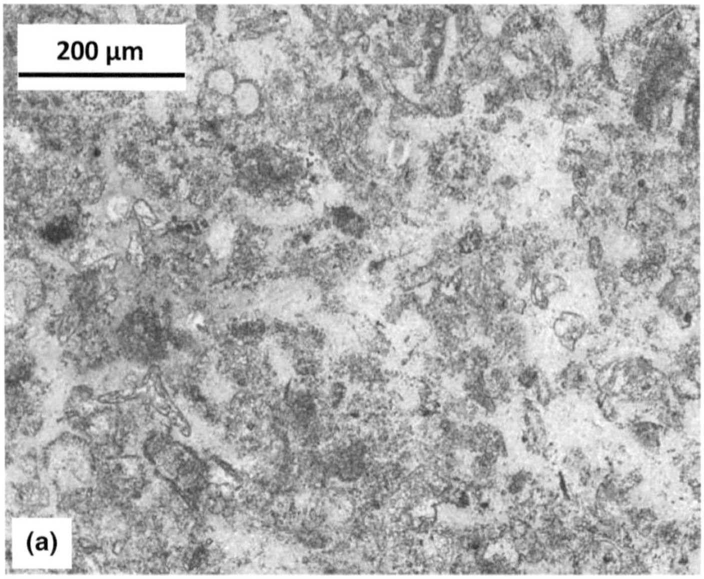

Chert (DK)

Chert composed of cryptocrystalline quartz, opaline silica, and chalcedony, with trace amounts of carbonate minerals (a, PPL). Local reddish-brown limonite-staining is from altered iron-containing minerals. The siliceous and/or silicified material comprises various types of foraminifera, sponge spiculae, and fossil debris (bioclasts). Some fossil remnants consist of carbonate (b, XPL); others are filled with feathery chalcedony (c, XPL). The matrix is mainly composed of cryptocrystalline quartz, with interstitial (optically isotropic) opaline silica.

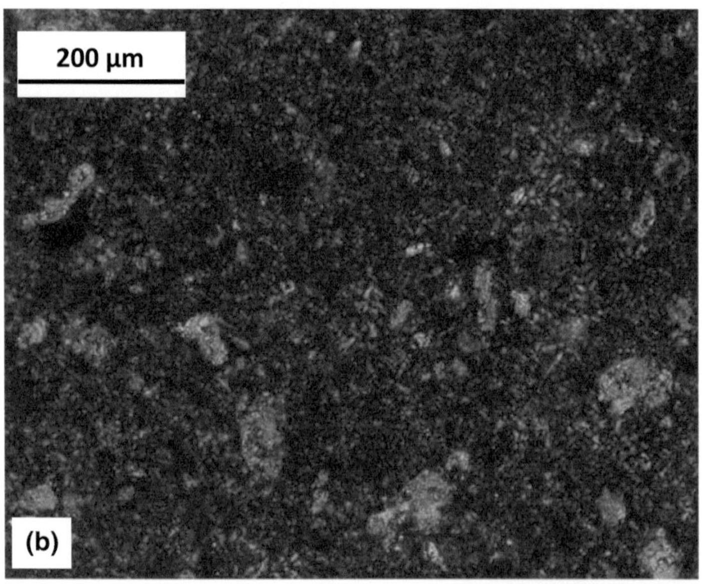

Reactivity potential attributed to:
Cryptocrystalline quartz, opaline silica and chalcedony. This aggregate may present pessimum behaviour (according to literature).

Aggregate classified as potentially reactive in laboratory tests and in field structures and edifices (Denmark).

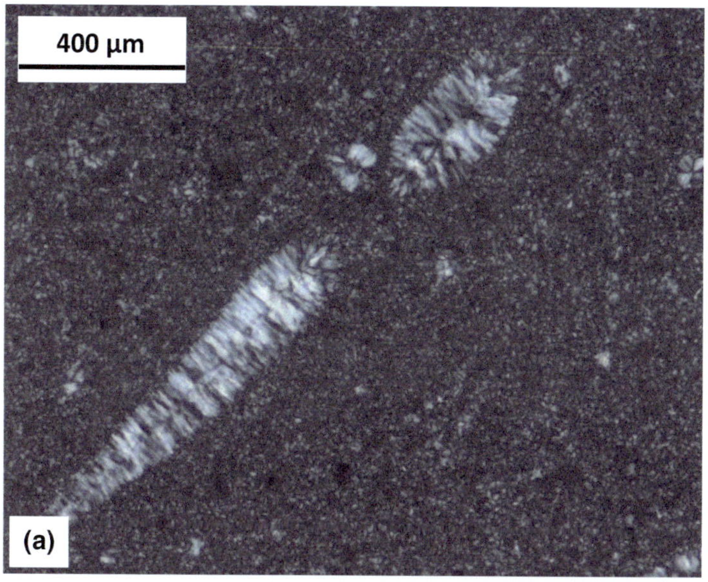

Chert (FR)

Chert composed of cryptocrystalline quartz, opaline silica, and chalcedony (a, XPL). It contains various foraminifera and fossil debris, filled with feathery chalcedony crystals perpendicular to the walls (b, XPL), replacing original carbonate material. The matrix is composed of very-fine grained cryptocrystalline quartz. The original aragonite of the shells was replaced by opal or chalcedony. The rock is internally micro-porous, larger voids are filled with fibrous chalcedony (c, XPL).

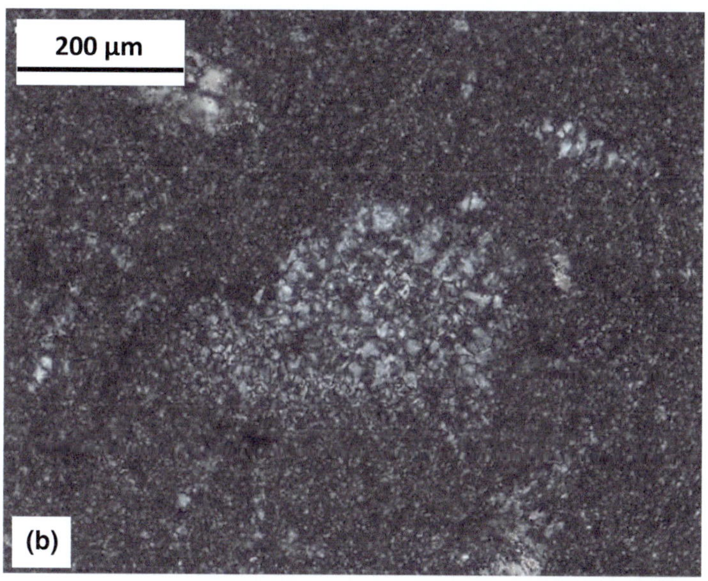

Reactivity potential attributed to:
Cryptocrystalline quartz, opaline silica and chalcedony.

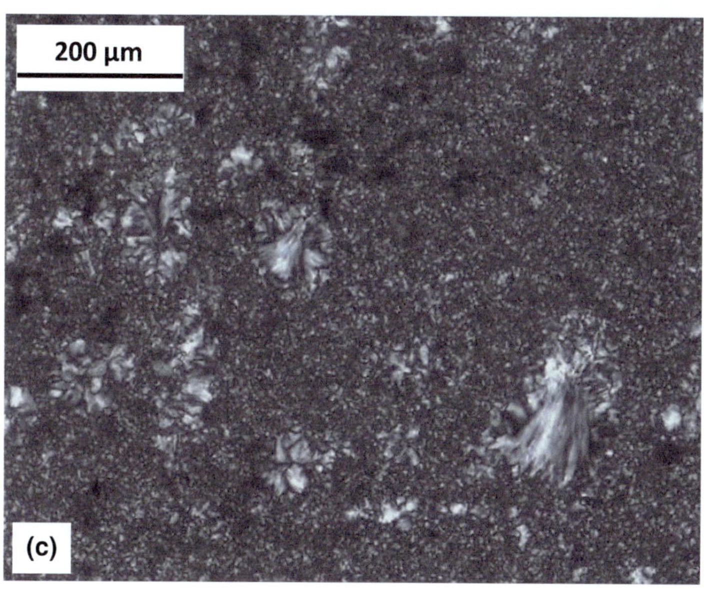

Lithology known to develop deleterious ASR in bridges and water structures (France).

Chert (FR)

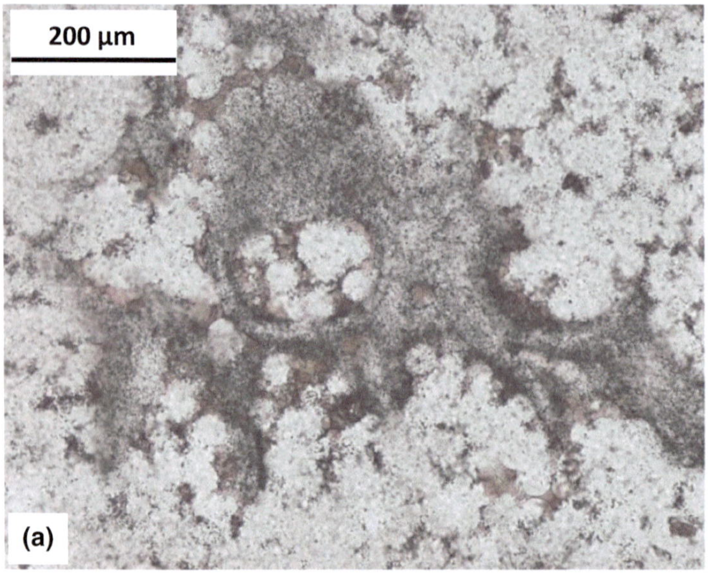

Radiolarian chert composed of agglomerated fossil radiolarian, consisting of microcrystalline and cryptocrystalline quartz, opaline silica, and chalcedony, locally stained reddish-brown with iron (oxy-) hydroxides from altered iron minerals (a, PPL). Spherical radiolarian fossils forming the bulk of the material are readily recognized, and are recrystallized to form very-fine grained cryptocrystalline quartz (b, GPXPL). Microcrystalline quartz is seen to fill in voids between fossil relics, together with minor fibrous chalcedony (c, XPL).

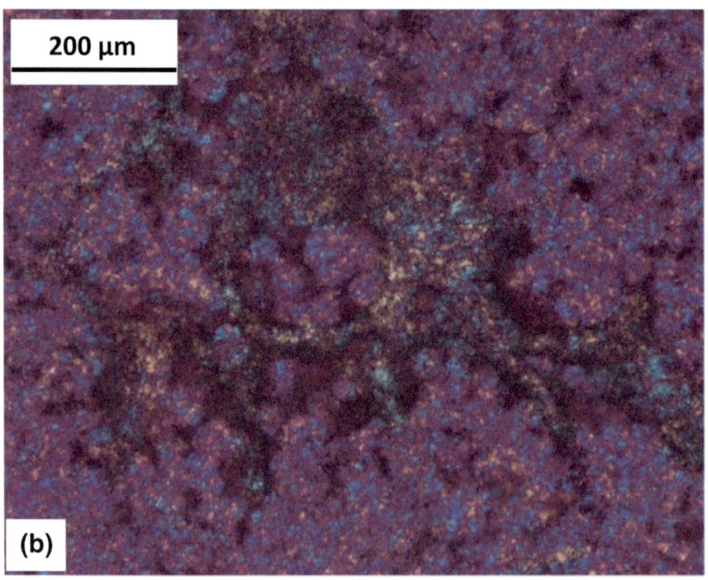

Reactivity potential attributed to:
Microcrystalline and cryptocrystalline quartz, opaline silica and chalcedony.

Lithology known to develop deleterious ASR in bridges and water structures (France).

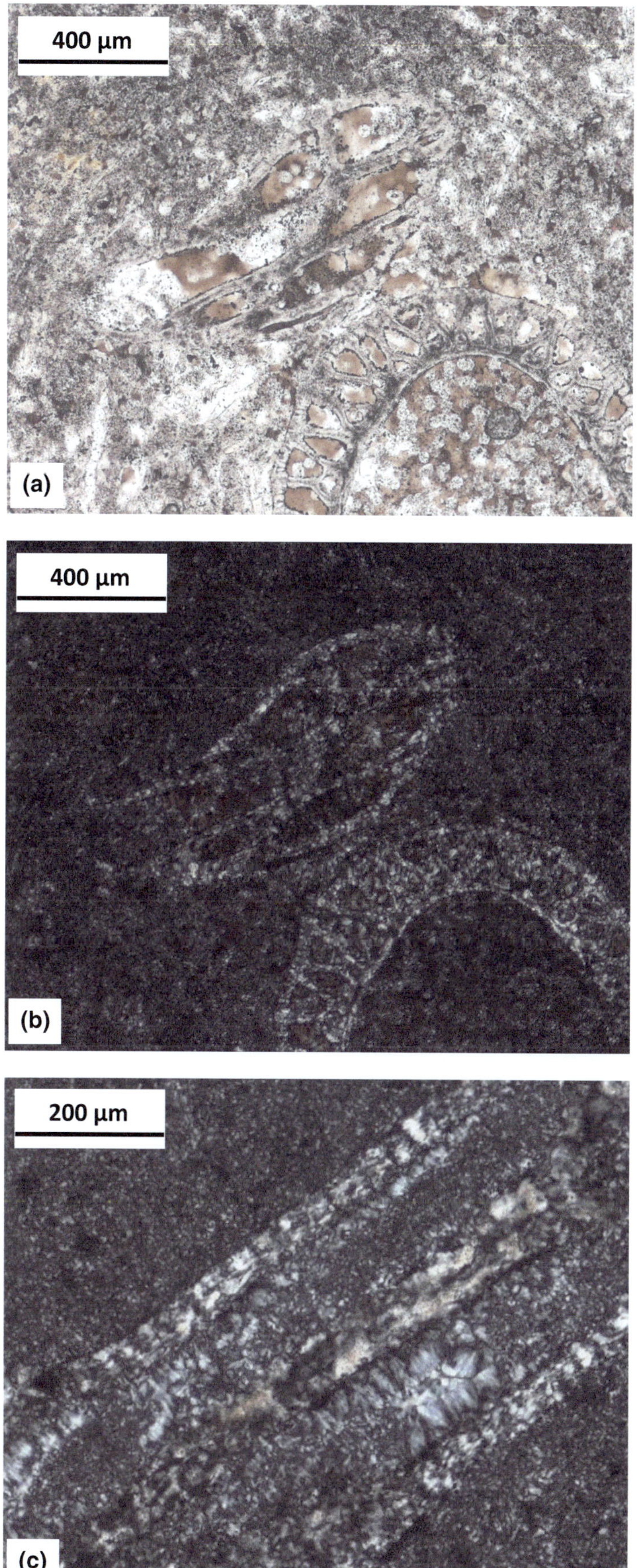

Chert (FR)

Fossiliferous chert composed of bioclasts, filled with microcrystalline quartz and opaline silica stained brownish by iron (oxy-) hydroxides (a, PPL). The matrix is composed of very-fine grained cryptocrystalline quartz and opaline silica (b, XPL). Fibrous chalcedony fills voids within bioclasts (c, XPL). The multicellular fossil relics are many size orders coarser (i.e. millimetre) than the grain size of the composing silica varieties (micrometre).

Reactivity potential attributed to:
Microcrystalline and cryptocrystalline quartz, opaline silica and chalcedony.

Lithology known to develop deleterious ASR in bridges and water structures (France).

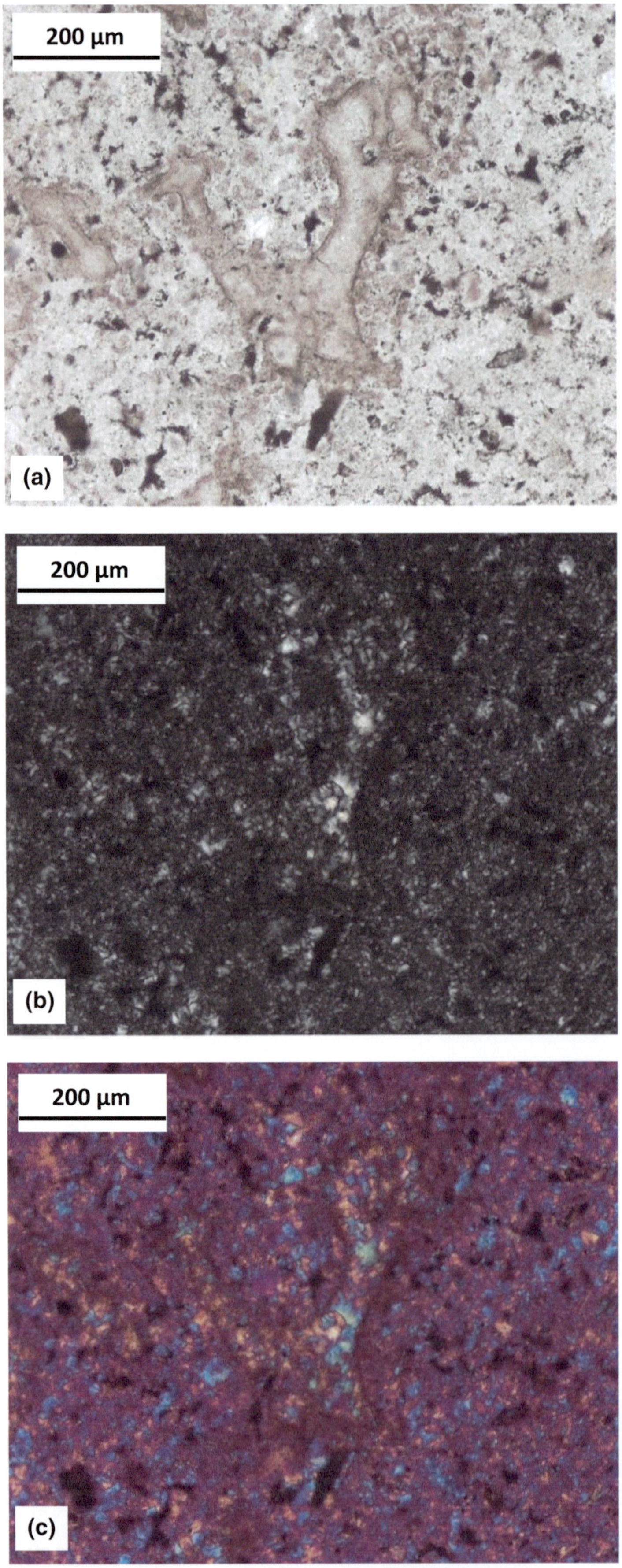

Chert (FR)

Chert composed of cryptocrystalline quartz, chalcedony and opaline silica, locally stained reddish-brown by iron (oxy-) hydroxides (limonite), with finely interspersed organic matter (a, PPL). Radiolarian fossils are recognized by their round shape, as well as sparsely distributed foraminifera and fossil fragments (bioclasts). Interstitial voids and the interior of larger fossils are filled with brownish stained feathery chalcedony. The fossil remains themselves are often replaced with microcrystalline quartz (b, XPL). The matrix is dominated by cryptocrystalline quartz, locally with interstitial isotropic opaline silica appearing burgundy red (c, GPXPL).

Reactivity potential attributed to:
Microcrystalline and cryptocrystalline quartz, opaline silica and chalcedony.

Lithology known to develop deleterious ASR in bridges and water structures (France).

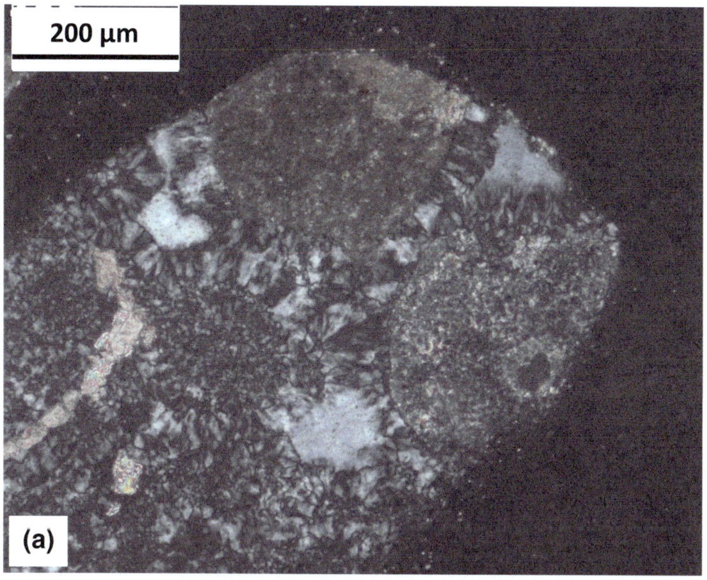

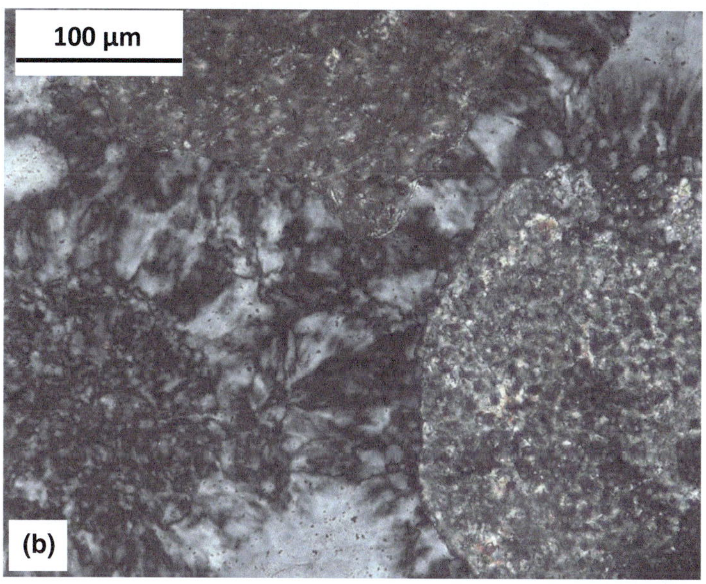

Chert (FR)

Fossiliferous chert containing cryptocrystalline quartz, chalcedony and trace amounts of calcite (a, XPL). The rock is composed of abundant originally carbonate fossil fragments (bioclasts e.g. various types of foraminifera), replaced with cryptocrystalline quartz with minor carbonate relics (b, XPL). Interclast voids are filled with feathery chalcedony, locally recrystallized to microcrystalline quartz (c, XPL).

Reactivity potential attributed to:
Microcrystalline and cryptocrystalline quartz and chalcedony.

Lithology known to develop deleterious ASR in bridges and water structures (France).

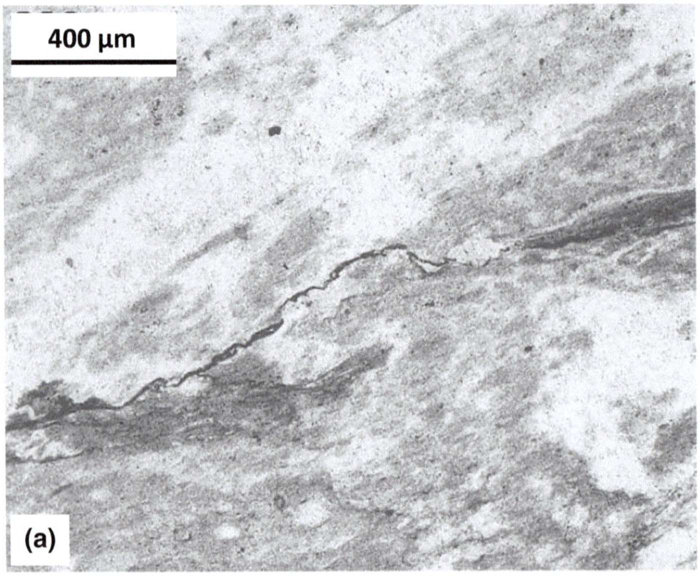

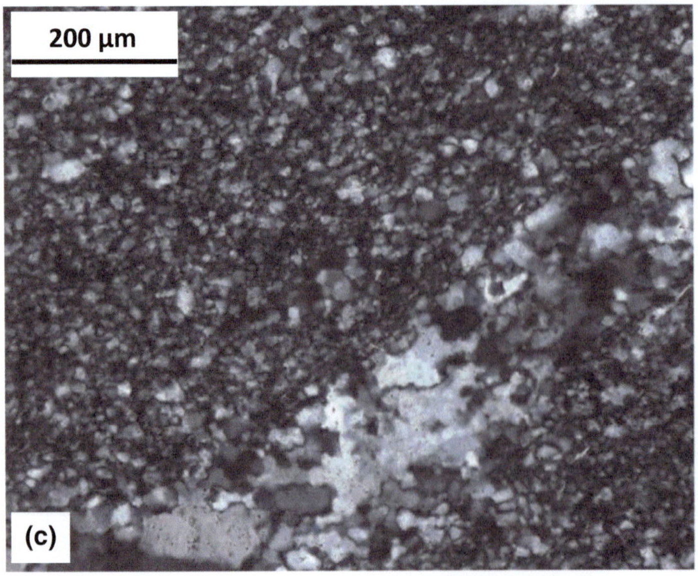

Chert (FR)

Very-fine grained siliceous rock composed of microcrystalline to cryptocrystalline quartz, stained by red to brownish iron oxides, mostly hematite (a, PPL). This rock type is also known as jasper, and may appear of ochre to brick red in various tones in hand specimen. Cryptocrystalline silica forms the bulk of the rock, microcrystalline quartz occurs in veinlets and smaller domains (b, XPL). Pressure-solution features occur throughout the section (a, PPL), lined by clay minerals and possibly limonite. The cryptocrystalline quartz matrix is locally polygonal (b, XPL), and microcrystalline quartz in veinlets often exhibits serrate grain boundaries (c, XPL).

Reactivity potential attributed to:
Microcrystalline and cryptocrystalline quartz.

Lithology known to develop deleterious ASR in bridges and water structures (France).

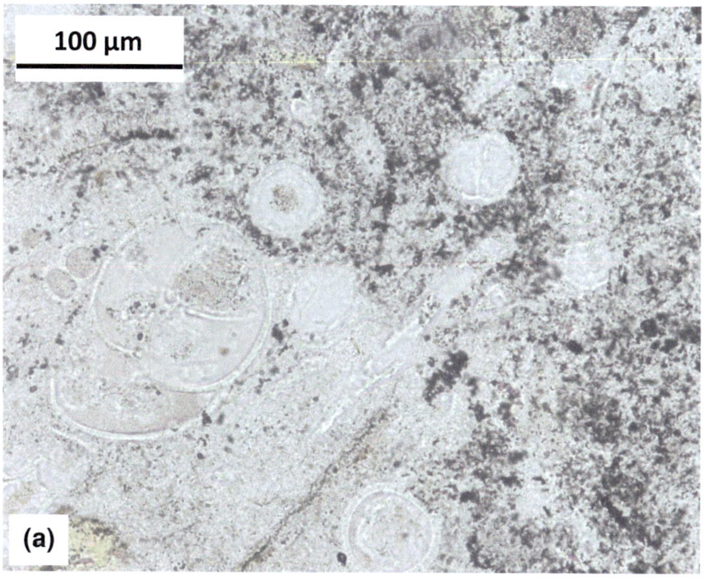

Chert (NL)

Chert containing cryptocrystalline quartz, opaline silica, chalcedony, and traces of clay minerals (a, PPL). Whereas mineral content is rather consistent, texture, fossil content, internal porosity, presence of interstitial clay minerals and/or sulfide minerals varies among and across individual chert particles. Foraminifera and elongate sponge spiculae can be discerned (b, XPL). Lighter areas in PPL represent comparatively large crystals of quartz. The interior of fossil relics is often filled with feathery chalcedony and opaline silica (c, XPL).

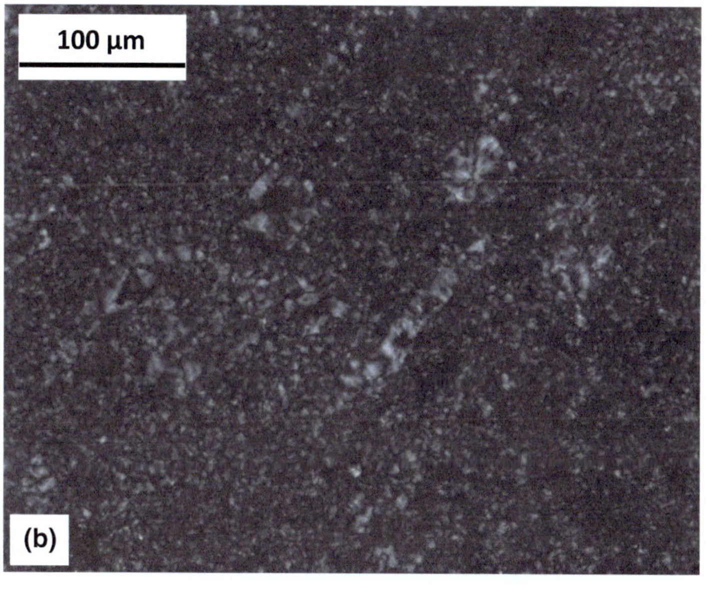

Note: this section was impregnated with a yellow-dyed epoxy for fluorescence.

Reactivity potential attributed to:
Cryptocrystalline silica, opaline silica and chalcedony.

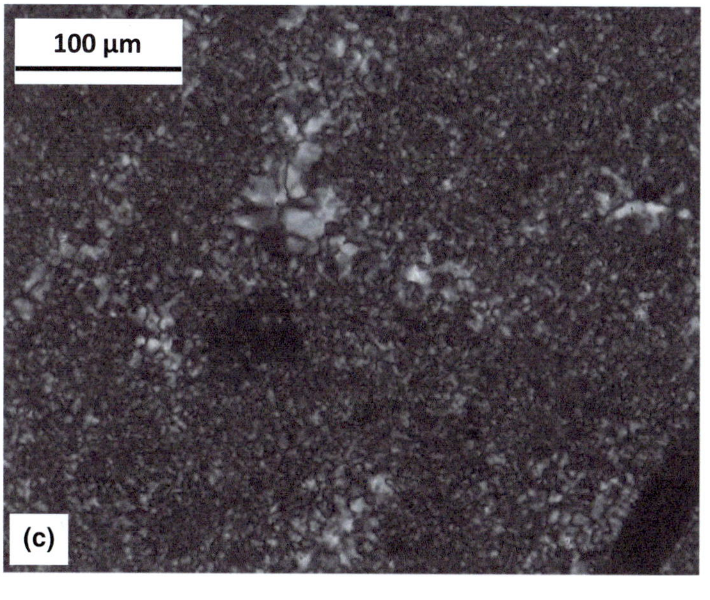

Aggregate particle in a sample of concrete from a viaduct *(The Netherlands)*.

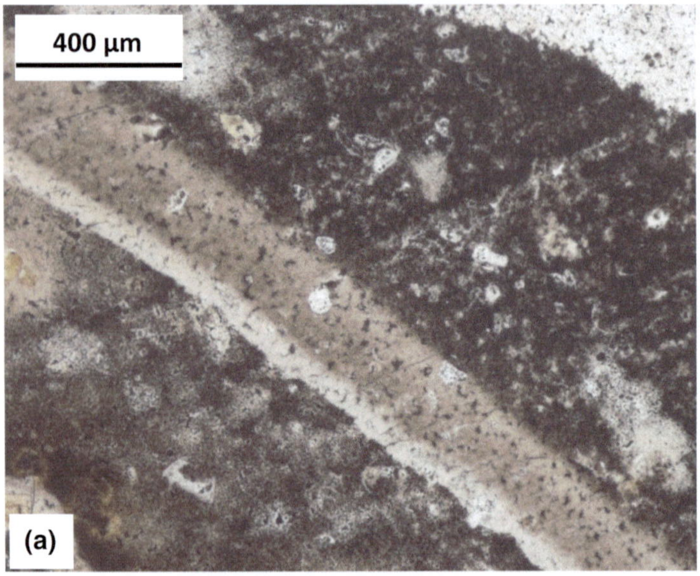

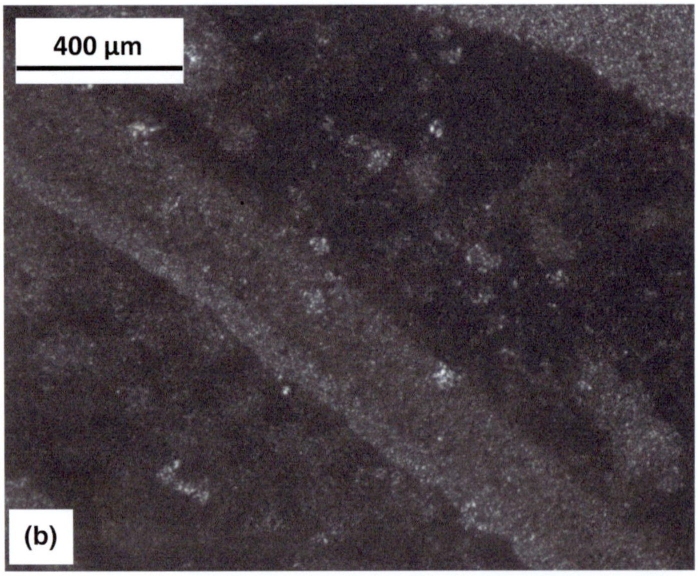

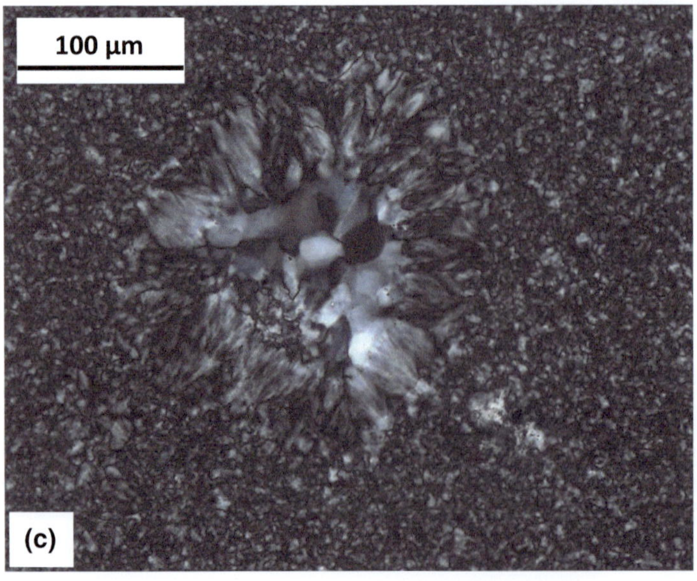

Chert (NL)

Porous chert with diatom fossils, containing opaline silica and cryptocrystalline quartz, stained brown by impregnation with iron (oxy-) hydroxides (limonite) (a, PPL). Porosity is higher in the lighter coloured cortex than in the interior (b, XPL). Diatom fossils are recognizable as light-coloured pockets lined with feathery chalcedony, embedded in an extremely fine-grained matrix of cryptocrystalline quartz (c, XPL).

Note: this section was impregnated with a yellow-dyed epoxy for fluorescence.

Reactivity potential attributed to:
Cryptocrystalline quartz, opaline silica and chalcedony.

Aggregate particle of the sand fraction in a sample of concrete from a viaduct (The Netherlands).

Chert (UK)

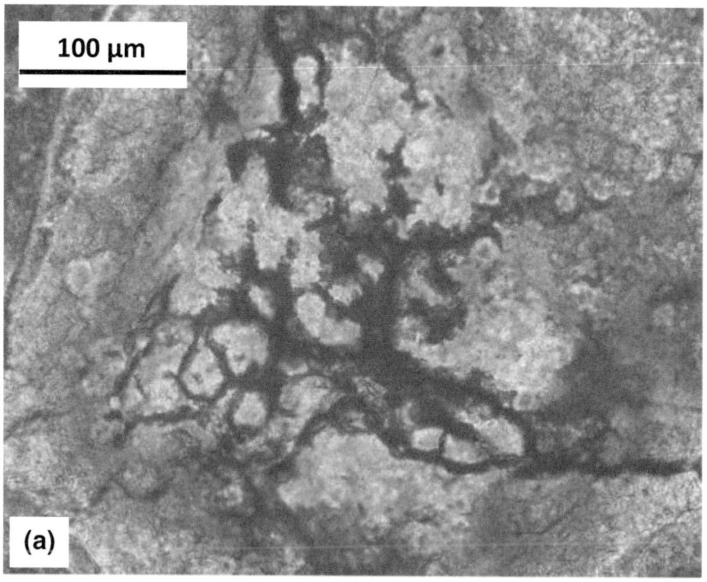

Non-porous, dense chert composed of cryptocrystalline silica, stained with reddish-brownish iron (oxy-) hydroxides (limonite) filling internal cracks (a, PPL). Some particles may show partial development of a cortex with porosity different from the interior (both more and less porous can be observed), and slightly coarser grain size (b, XPL). Due to the very-fine grained nature of the chert, it is not possible to determine individual mineral grains, but cryptocrystalline quartz and chalcedony are identifiable using the gypsum plate (c, GPXPL).

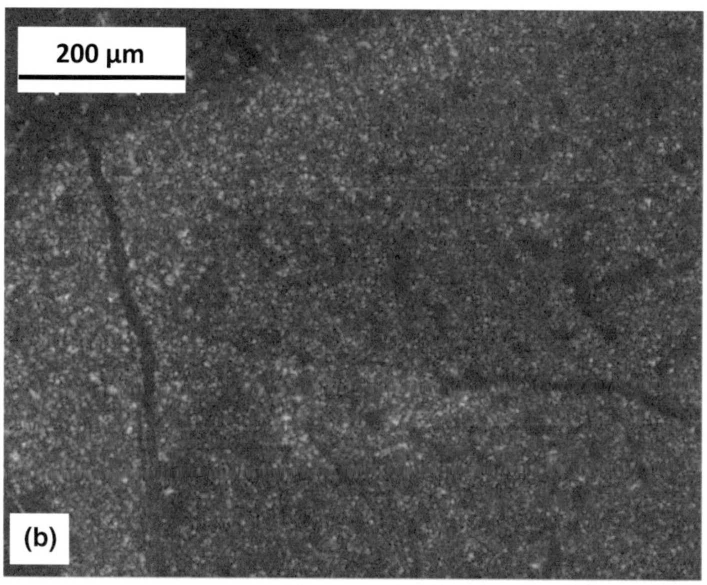

Reactivity potential attributed to:
Cryptocrystalline quartz, opaline silica and chalcedony.

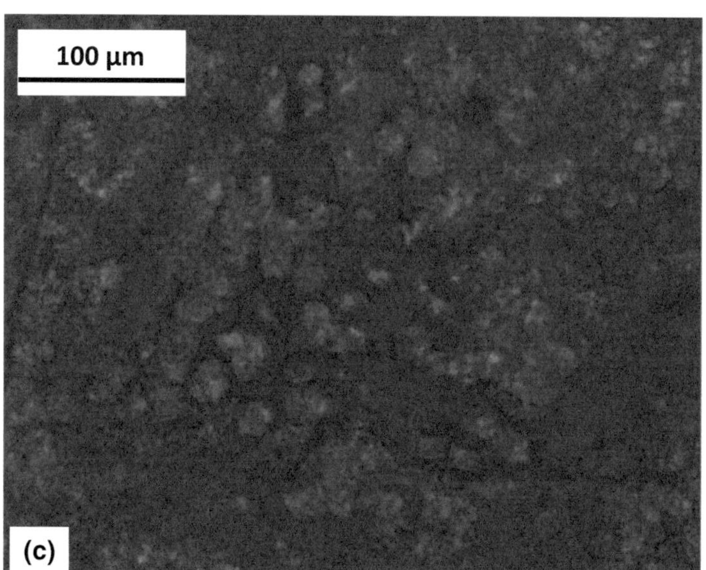

Aggregate particle in a sample of concrete from a bridge (United Kingdom).

Greywacke (AU)

Medium-grained sedimentary rock composed of quartz, K-feldspar, plagioclase, with minor lithic fragments (chert), and chlorite, trace amounts of sericite, apatite, zircon, and calcite of secondary origin (a, XPL). Clasts are sub-angular to rounded and show wide size variation. K-feldspar and plagioclase appear cloudy from advanced alteration (b, XPL). Matrix is abundant and contains cryptocrystalline quartz, sericite, chlorite and calcite. Cryptocrystalline quartz is present both in clasts and in matrix recrystallized under diagenetic to very-low grade metamorphic conditions (c, XPL).

Reactivity potential attributed to:
Microcrystalline to cryptocrystalline quartz.

Aggregate particle in a concrete prism (Australia).

Greywacke (CA)

Fine-grained greywacke composed of coarse angular clasts of quartz, lithic fragments (metamorphicrock), plagioclase, K-feldspar, biotite, minor muscovite, with trace amounts of chlorite, organic matter, titanite, calcite and zircon. The matrix is composed of abundant cryptocrystalline quartz, sericite and chlorite (a, XPL). Lithic fragments contain strained microcrystalline metamorphic quartz. Partly chloritised biotite (b, XPL) seem aligned to a weakly developed preferred orientation. The absence of undulatory extinction in quartz, as well as preserved straight biotite flakes (b, XPL) and polysynthetic plagioclase twins (c, XPL) indicates negligible diagenetic deformation. Fine grained sericite in the matrix is of neogenic origin formed under diagenetic to very low-grade metamorphic conditions (c, XPL).

Reactivity potential attributed to:
Microcrystalline to cryptocrystalline in the matrix and strained quartz in the clasts.

Aggregate particle in a sample of concrete prism (Canada).

Greywacke (CA)

Medium-grained greywacke composed of sub-angular to sub-rounded clasts of quartz, K-feldspar, plagioclase, muscovite, with minor amounts of sericite, pyroxene, biotite, chlorite, and traces of zircon, epidote, titanite, and rutile. The fine-grained matrix contains microcrystalline quartz, sericite, biotite partially altered to chlorite (a, XPL). Grain contacts are concave-convex to serrate (b, XPL). Large clasts do not reveal any significant diagenetic deformation. Elongate or flaky minerals in the matrix define a poorly developed preferred orientation of sedimentary origin (c, XPL). The rock matrix is cemented with microcrystalline to cryptocrystalline quartz and sericite of authigenic origin, formed under diagenetic to very low-grade metamorphic conditions.

Reactivity potential attributed to:
Microcrystalline to cryptocrystalline quartz in the matrix.

Sample of rock classified as potentially reactive in laboratory tests (Canada).

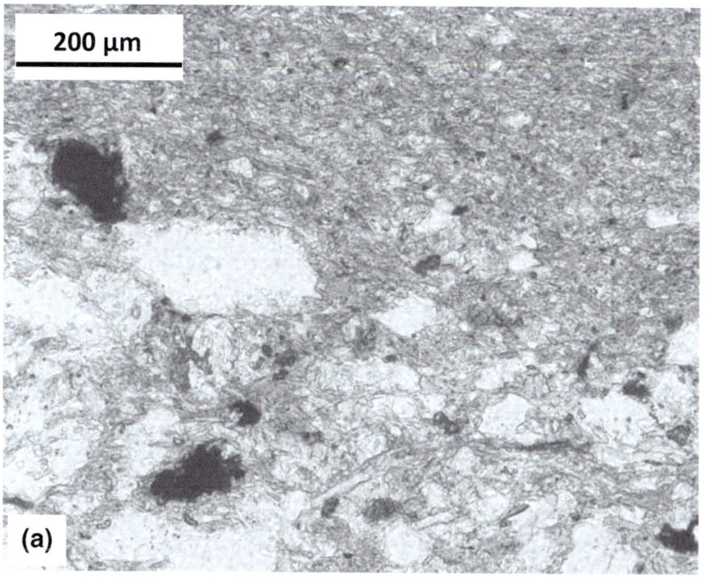

Greywacke (CA)

Coarse to very fine-grained greywacke composed of quartz, K-feldspar, polycrystalline siliceous lithic fragments, plagioclase, muscovite, chlorite, with a fine-grained matrix consisting of sericite, chlorite, organic matter and opaque minerals. The rock displays layering of grain-supported coarse clasts alternating with fine-grained mud (a, PPL), in which elongate and flaky minerals display a preferred orientation (b, XPL). Opaque minerals may represent organic matter matured during compaction and consolidation. K-feldspar and plagioclase may appear cloudy from partial alteration. Large clasts show negligible undulatory extinction, feldspar twins and mica flakes are preserved straight, together indicating negligible diagenetic deformation (b, XPL). The matrix contains authigenic quartz and sericite formed under diagenetic conditions (c, XPL).

Reactivity potential attributed to:
Microcrystalline to cryptocrystalline quartz in the matrix of the greywacke level and in the mudstone level.

Aggregate particle in a sample of concrete prism (Canada).

Greywacke (UK)

Greywacke containing subrounded to angular coarse to medium-grained clasts of quartz and lithic fragments (e.g. chert, siltstone), and minor plagioclase, muscovite, chlorite (a, PPL). The matrix comprises microcrystalline to cryptocrystalline quartz and sericite of prograde neogenic origin (b, XPL). The dusty appearance in PPL is caused by clusters of organic matter. The internal fabric of lithic fragments with microcrystalline to cryptocrystalline quartz indicates a low-grade metamorphic provenance (b, XPL). Large quartz clasts display uniform extinction, and feldspar twin lamellae and mica flakes are straight (c, XPL), suggesting that deformation from load-compaction is negligible. Clasts appear cemented by amorphous low-relief non-birefringent material (authigenic opaline silica?) deposited along the periphery of clastic particles.

Note: this section was impregnated with a yellow-dyed epoxy for fluorescence.

Reactivity potential attribute d to:
Microcrystalline and cryptocrystalline quartz in the matrix and in clasts.

Sample of rock classified as potentially reactive in laboratory tests and also in concrete structures such as dams, bridges and concrete road pavements (United Kingdom).

Greywacke (USA)

Medium-grained lithic greywacke, composed of lithic fragments (e.g. slate, shale, chert), and quartz clasts, minor amounts of plagioclase, K-feldspar, and traces of muscovite, sericite, chlorite, titanite, calcite and clay minerals forming the fine-grained matrix interspersed with organic matter (a, XPL). K-feldspar and plagioclase may appear cloudy from partial alteration to kaolinite/sericite. Coarse, subrounded to angular, occasionally elongate lithic fragments of sedimentary or diagenetic provenance are predominant (b, XPL). Grain contacts are locally concave-convex, and may be lined with neogenic sericite from prograde alteration of clay under diagenetic or very low-grade metamorphic conditions (c, XPL).

Reactivity potential attributed to:
Microcrystalline to cryptocrystalline quartz in lithic fragments and matrix.

Sample of rock classified as potentially reactive in laboratory tests (USA).

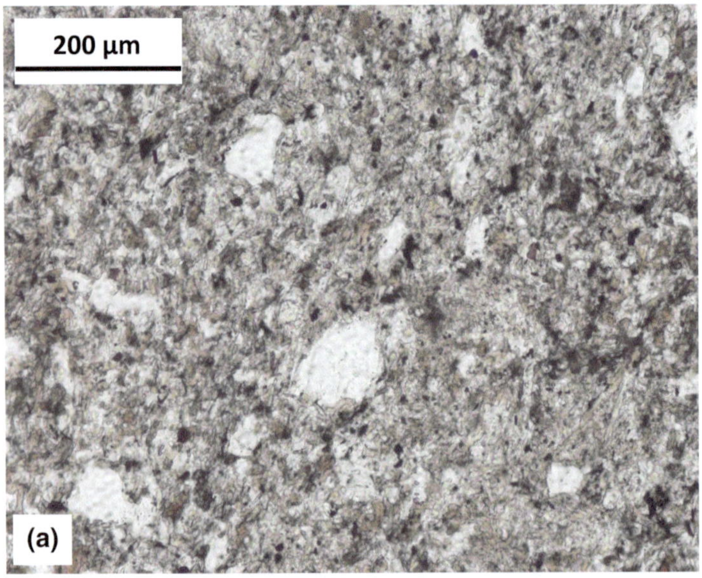

Greywacke (ZA)

Fine-grained rock composed of angular clasts of comparatively coarser quartz, plagioclase and K-feldspar clasts embedded in a fine-grained matrix of microcrystalline to cryptocrystalline quartz, biotite (red-brown and green), sericite, and opaque minerals (a, PPL). Elongate and flaky clasts exhibit a weakly developed preferred orientation of sedimentary origin (b, XPL). The lack of deformation features in the coarser clasts fraction indicates minimal compaction-deformation (c, XPL). Part of the quartz and sericite in the matrix may be of authigenic origin, formed under diagenetic to very low-grade metamorphic conditions.

Reactivity potential attributed to:
Microcrystalline to cryptocrystalline quartz in the matrix.

Aggregate particle in a concrete sample (South Africa).

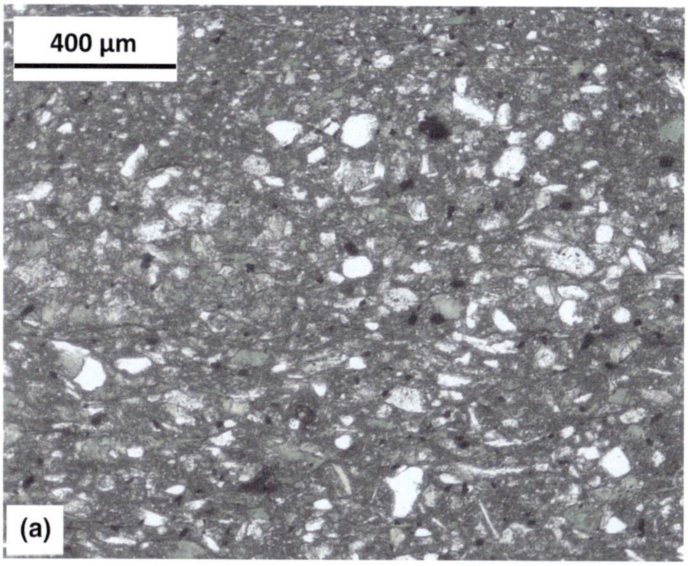

Greywacke (ZA)

Fine-grained greywacke composed of angular to sub-rounded clasts of quartz, chlorite, and polycrystalline calcite, with minor plagioclase and muscovite. The matrix is very-fine grained and contains microcrystalline to cryptocrystalline quartz, sericite, carbonate, accessorial zircon, and opaque specks (a, PPL). Carbonate clasts are of detrital origin, but neogenic carbonate is seen to fill voids (b, XPL). Also very-fine grained quartz, sericite and chlorite in the matrix may be of neogenic origin, formed under prograde diagenetic conditions (c, XPL).

Reactivity potential attributed to:
Microcrystalline and cryptocrystalline quartz.

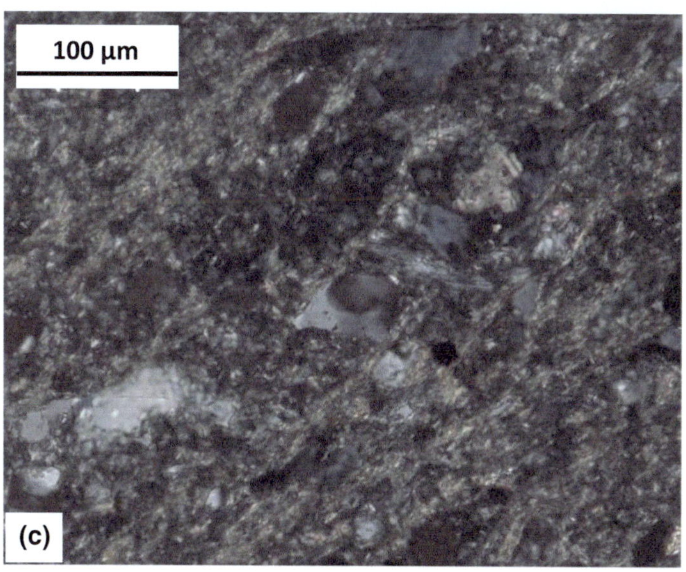

Aggregate particle in concrete from a retaining wall (South Africa).

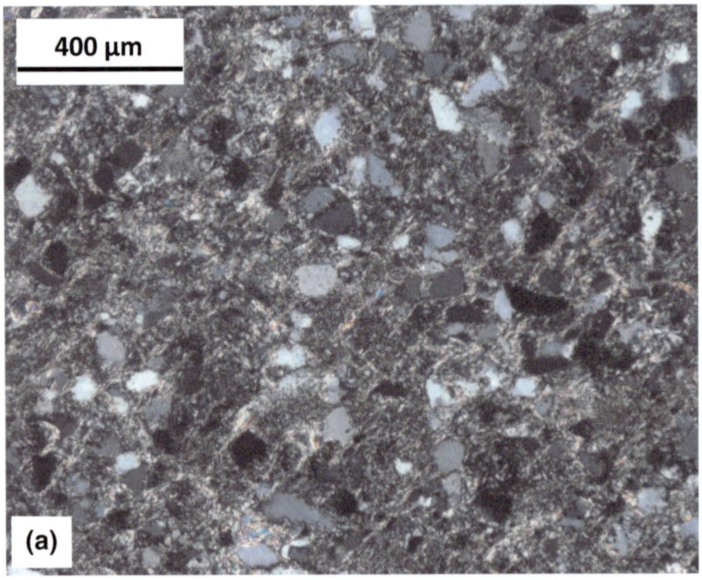

Greywacke (ZA)

Fine-grained clastic rock composed of angular quartz, lithic fragments (quartzite), muscovite/sericite, and trace amounts of biotite, plagioclase, K-feldspar, tourmaline, titanite, and interspersed organic matter. The abundant very fine-grained siliceous matrix contains microcrystalline quartz and sericite that may be partially prograde of diagenetic origin (a, XPL). Plagioclase and K-feldspar appear cloudy from partial alteration to sericite or kaolinite, respectively (b, XPL); biotite is partially altered to chlorite. Flaky minerals in the matrix may locally show a poorly developed preferred orientation (c, XPL).

Reactivity potential attributed to:
Microcrystalline quartz in the matrix and in some lithic fragments.

Aggregate particle in concrete from a retaining wall (South Africa).

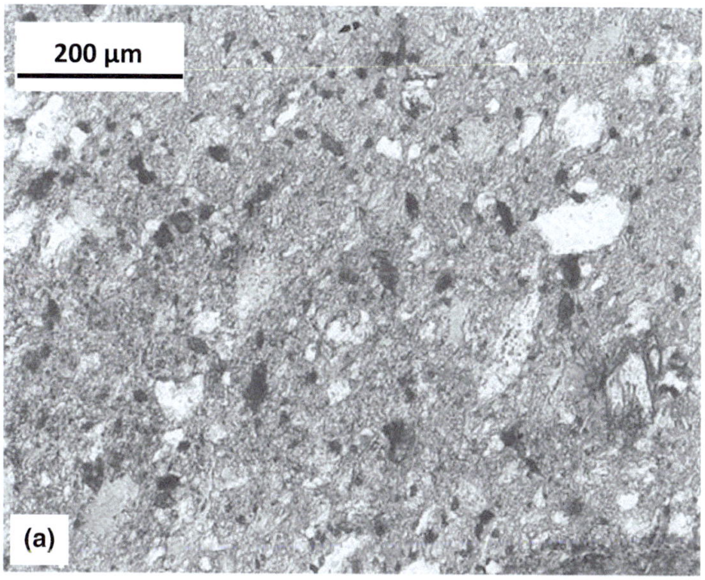

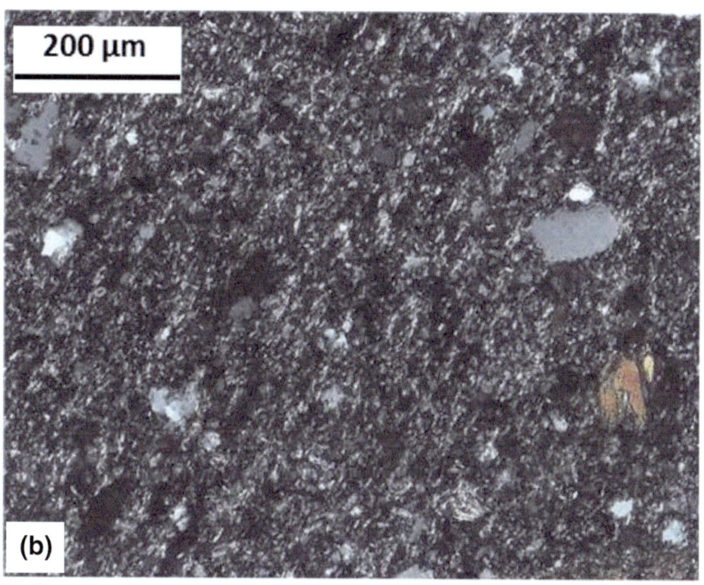

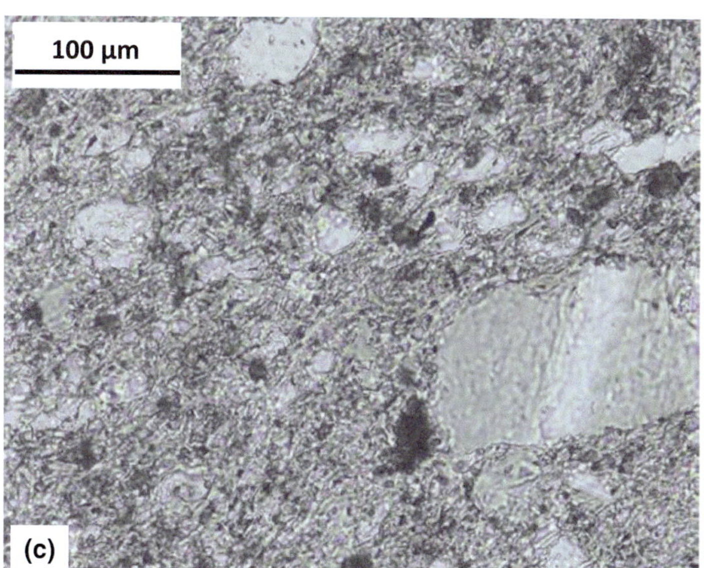

Greywacke (ZA)

Medium to very-fine grained rock composed of relatively coarse clasts of quartz, chlorite, muscovite, and epidote, embedded in a very-fine grained matrix of the same minerals, plus titanite and opaque minerals (a, PPL). Elongate and flaky minerals (micas) show a preferred orientation. The matrix contains abundant microcrystalline to cryptocrystalline quartz (b, XPL). Due to the very-fine grained nature of the matrix, particle size, shape, and mineral identity is not always clear. The neogenic chlorite is formed at low-grade metamorphic conditions (c, PPL).

Reactivity potential attributed to:
Microcrystalline to cryptocrystalline quartz in the matrix.

Aggregate particle in concrete from a retaining wall (South Africa).

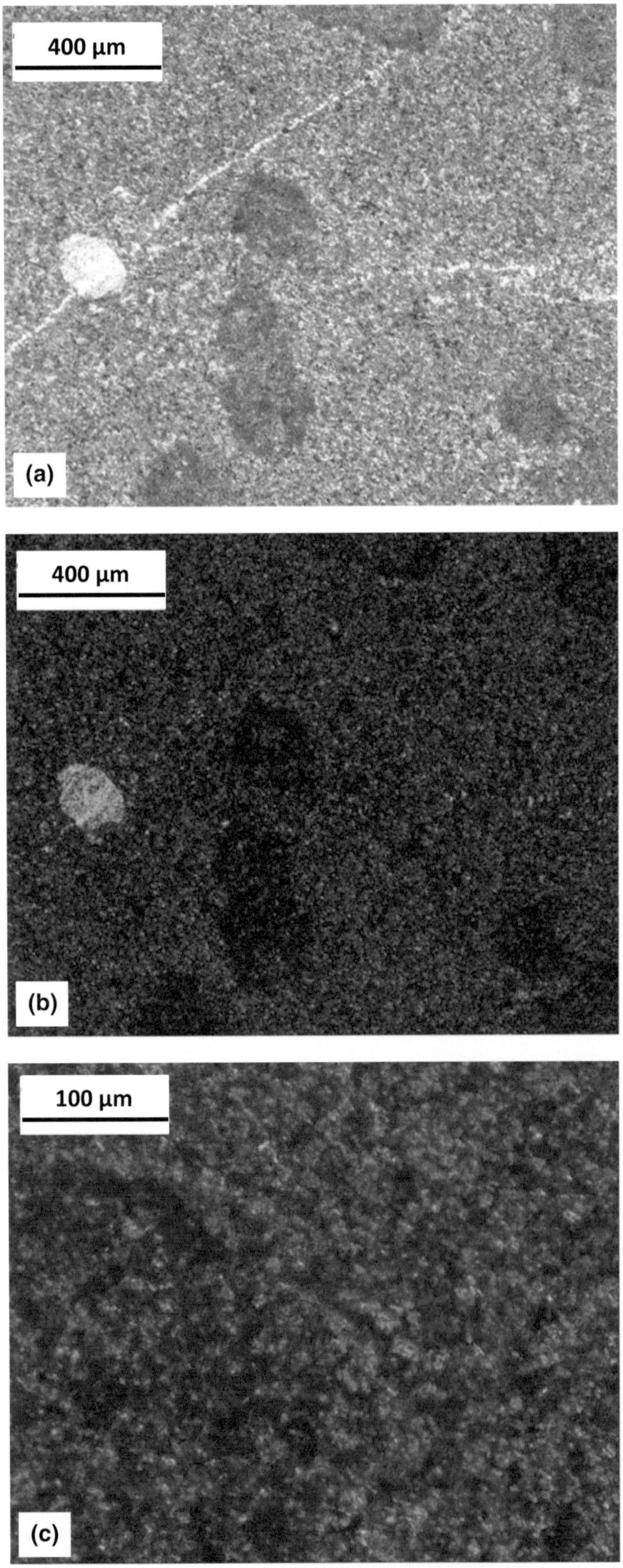

Mudstone (CA)

Very fine-grained mudstone composed of quartz, chlorite, glauconite, opaque minerals, clay minerals and microcrystalline quartz with rare clasts of K-feldspar (a, PPL) and plagioclase. The rock shows a poorly developed preferred orientation with nodules of glauconite dispersed and randomly oriented (a, PPL). In the matrix, microcrystalline to cryptocrystalline quartz occurs interlayered with the phyllosilicates, glauconite and chlorite (b, XPL). Under higher magnification the porous character of the glauconite is visible (c, XPL).

Reactivity potential attributed to:
Microcrystalline to cryptocrystalline quartz in the matrix.

Aggregate particle in a sample of concrete prism (Canada).

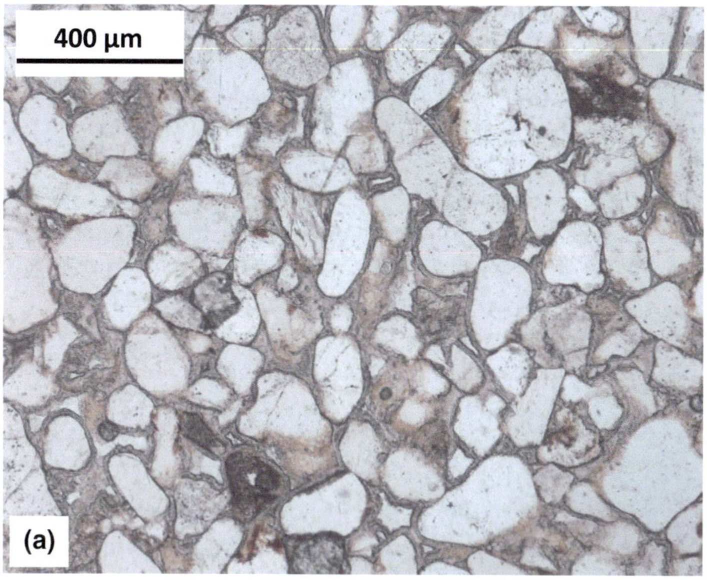

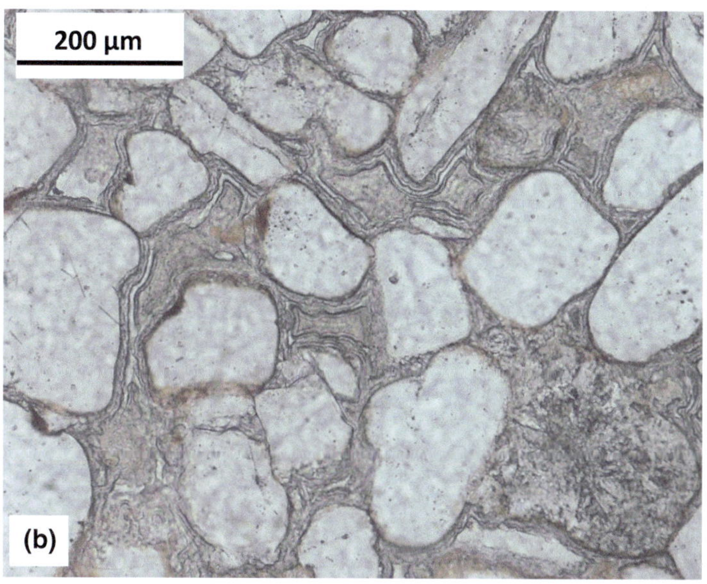

Sandstone (AR)

Well-sorted, medium-grained sandstone (quartz-arenite). Medium- to well-rounded clasts of mono-mineralic quartz and some feldspar, plus trace amounts of fragments of fine-grained lithologies (e.g. quartzite, sand-/siltstone, basalt), are cemented by reddish siliceous cement (a, PPL) containing chalcedony, opal, and cryptocrystalline silica. The degree of compaction is low, grain contacts are mostly point-shaped (a, PPL), a few oblong or slightly convex, but none serrate. Re-entrant parts along a clast perimeter are locally lined with reddish-brown iron (oxy-) hydroxides (limonite) and/or argillaceous material (b, PPL). Percussion cracks are occasionally visible in the particle interior (c, XPL). Locally, voids and pores are lined with chalcedony and opal (b, PPL) with a fibrous texture perpendicular to the interstice wall. Interstitial space is filled with plumiform chalcedony and isotropic amorphous silica (opal) (c, XPL).

Reactivity potential attributed to:
Cryptocrystalline silica in the interclastic cement, chalcedony and opaline silica (based on information in literature).

Sample of rock classified as potentially reactive in laboratory tests (Argentina).

Sandstone (AU)

Well-sorted, medium-grained sandstone composed of detrital quartz, minor plagioclase, K-feldspar, and trace amounts of tourmaline, titanite, opaque minerals and siliceous lithic fragments (chert). This is a mineralogically mature grain supported rock in which monomineralic quartz clasts are predominant (a, XPL). The sandstone is slightly deformed from load-compaction, with oblong to convex grain contacts. The original well-rounded particle shape is mostly obstructed by authigenic quartz growth in optical continuity (b, XPL), and interstitial spaces are occupied by microcrystalline quartz and sericite of neogenic origin. Feldspar clasts appear cloudy due to alteration to sericite or kaolinite. Lithic fragments of sedimentary provenance contain microcrystalline to cryptocrystalline quartz (c, XPL)

Reactivity potential attributed to:
Microcrystalline and cryptocrystalline quartz in lithic clasts and neogenic rock cement.

Aggregate particle in a sample of concrete (Australia).

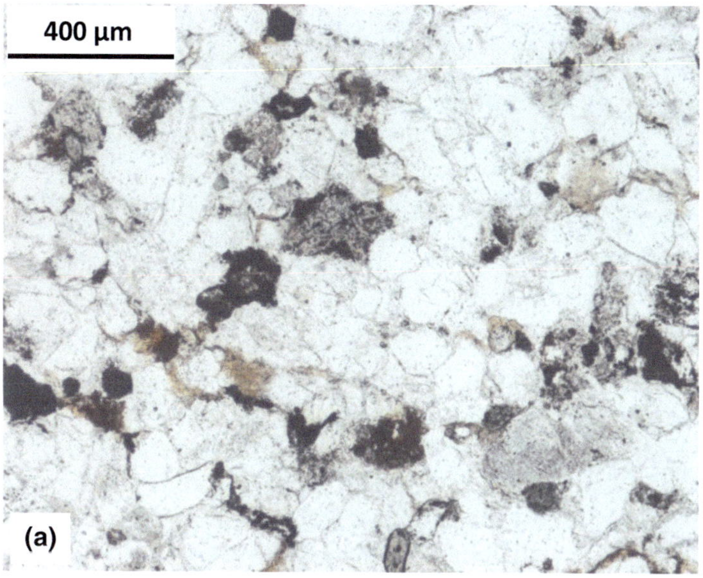

Sandstone (AU)

Medium to fine-grained moderately sorted sedimentary rock, mainly composed of monomineralic quartz particles, with minor feldspar and lithic fragments (chert), and trace amounts of sericite, tourmaline, opaque minerals, zircon, and limonite (a, PPL). Clasts are angular to subrounded. Locally serrate to interlobate grain boundaries are observed (b, XPL), lined by insoluble clay and/or opaque minerals. Authigenic silica overgrows monomineralic quartz clasts, revealing the original circumference lined by limonite (c, XPL).

Reactivity potential attributed to:
Microcrystalline quartz in lithic (chert) fragments, cryptocrystalline (authigenic) quartz cement.

Aggregate particle in a sample of concrete (Australia).

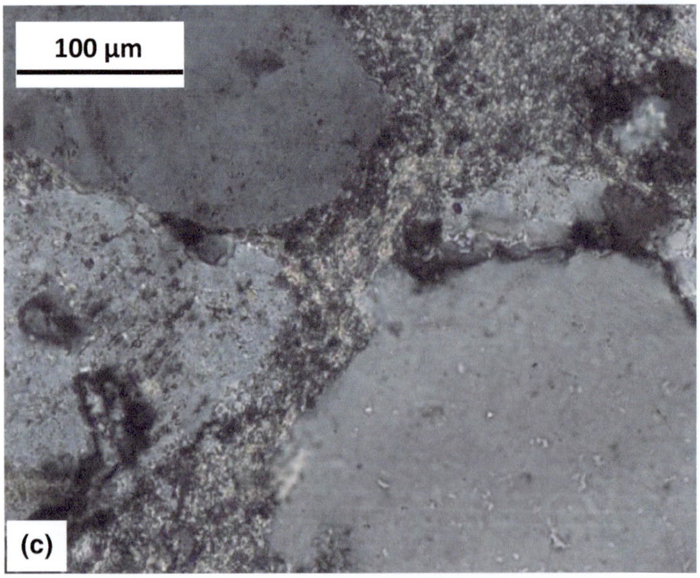

Sandstone (CA)

Clastic rock composed of clasts of well-rounded quartz, plagioclase, sub-rounded K-feldspar and lithic fragments (metamorphic), with subordinate interstitial fine-grained quartz-pelitic matrix. Monomineralic quartz clasts are predominant (a, XPL). K-feldspar grains are cloudy from alteration to kaolinite. The clasts have oblong shape with contacts developing into serrate f rom load-compaction, and may exhibit undulatory extinction, strain lamellae and subgrains (b, XPL). Original particle shape may be obscured due to localized pressure solution. The matrix is composed of microcrystalline to cryptocrystalline quartz and mica (sericite) (c, XPL).

Reactivity potential attributed to:
Microcrystalline to cryptocrystalline quartz, subgraining along boundaries, deformation lamellae.

Aggregate particle in a sample of concrete prism (Canada).

Sandstone (CA)

Poorly-sorted, fine-grained sandstone composed of subrounded to angular clasts of monomineralic quartz, plagioclase and lithic fragments, embedded by fine grained siliceous cement (a, XPL). Clay minerals, sericite, biotite and iron (oxy-) hydroxides are observed. Angular feldspar clasts are partially altered and corroded along the edges, overgrown by neogenic quartz (b, XPL). The interstitial cement contains microcrystalline to cryptocrystalline quartz and traces of sericite (c, XPL).

Reactivity potential attributed to:
Microcrystalline to cryptocrystalline quartz in the siliceous rock cement.

Aggregate particle in a sample of concrete prism (Canada).

Sandstone (CA)

Poorly-sorted sandstone composed of medium to coarse grained, angular and elongate clasts of quartz, plagioclase, K-feldspar, muscovite, biotite, and lithic fragments (quartzite), interspersed with finer-grained chlorite, organic matter, and titanite (a, XPL). The matrix in this grain-supported texture contains microcrystalline quartz, sericite, and biotite which is partly altered to chlorite (b, XPL). K-feldspar appears cloudy from partial alteration. Coarse clasts are mainly composed of monomineralic quartz. Metamorphic lithic fragments contain microcrystalline to cryptocrystalline quartz. Neogenic interstitial sericite may show a weakly developed preferred orientation. Adjacent quartz grains show lobated boundaries owing to recrystallization (c, XPL).

Reactivity potential attributed to:
Microcrystalline quartz in the matrix and strained quartz in lithic fragments.

Aggregate particle in a sample of concrete prism (Canada).

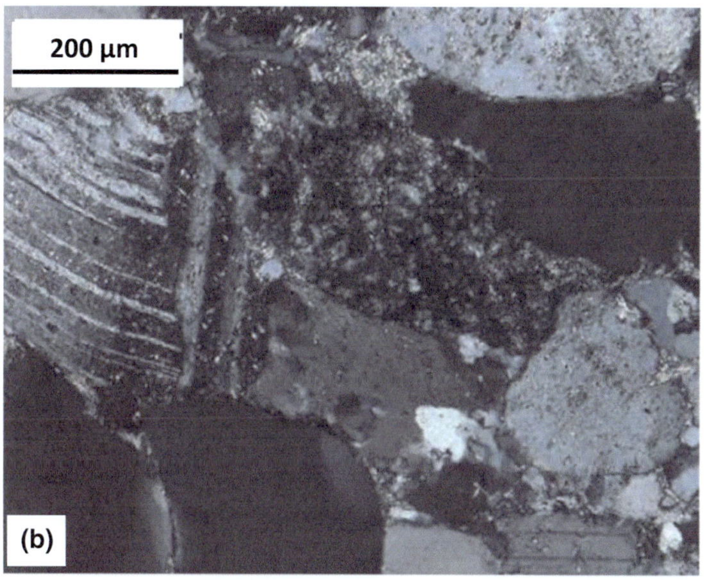

Sandstone (CA)

Moderately-sorted, lithic sandstone composed of sub-rounded to rounded clasts of metamorphic lithic fragments, quartz, K-feldspar and plagioclase, with sericite and cryptocrystalline quartz in the scarce matrix (a, XPL). Feldspars may show bent or kinked twinning planes from load-compaction, and grain contacts are oblong, straight to curved (b, XPL). Lithic fragments of fine-grained quartzite or chert are observed to contain microcrystalline to cryptocrystalline quartz (b, XPL). The sediment is cemented by minor authigenic quartz and sericite grown under diagenetic to very low-grade metamorphic conditions (c, XPL).

Reactivity potential attributed to:
Microcrystalline and cryptocrystalline quartz in the rock fragments. Cryptocrystalline quartz in the matrix.

Aggregate particle in a sample of concrete prism (Canada).

Sandstone (CN)

Poorly sorted sandstone composed of coarse detrital clasts of sub-rounded metamorphic lithic fragments and angular monomineralic quartz, finer particles of biotite, plagioclase and K-feldspar, chlorite, opaque minerals, and organic matter. The very fine-grained matrix consists of argillaceous mud coloured dark by organic matter (a, PPL). Cloudy plagioclase is partially altered (b, XPL). Lithic fragments represent polycrystalline quartzite with serrated grain boundaries, with biotite along clast edges. The rock shows a slight preferred orientation of elongate or flaky clasts (c, XPL). Load-compaction is minimal.

Reactivity potential attributed to:
Microcrystalline quartz in the matrix. The quartzite lithic fragments contain fine-grained quartz with serrated grain boundaries.

Sample of rock classified as potentially reactive in laboratory tests (China).

Sandstone (DE)

Poorly sorted sandstone composed of medium to coarse grained monomineralic quartz, K-feldspar, sericite as well as quartz-rich lithic fragments (metamorphic) (a, XPL). The original shape of the clastic particles is obscured by load-compaction and deformation. Deposition of subhedral authigenic quartz in optical continuity obliterates the original shape of the clasts (b, XPL). The neogenic quartz does not show fluid inclusion trails. Lithic fragments show polycrystalline quartz with sutured boundaries, bulging and subgrains (c, XPL). The rock shows cracks, voids and scarce rock cement.

Note: this section was impregnated with a yellow-dyed epoxy for fluorescence.

Reactivity potential attributed to:
Recrystallised quartz with sutured boundaries in the rock fragments (quartzite). Neogenic quartz. Porosity and cracks facilitate access of the pore solutions.

Aggregate particle in a sample of concrete from a hydraulic structure (Germany).

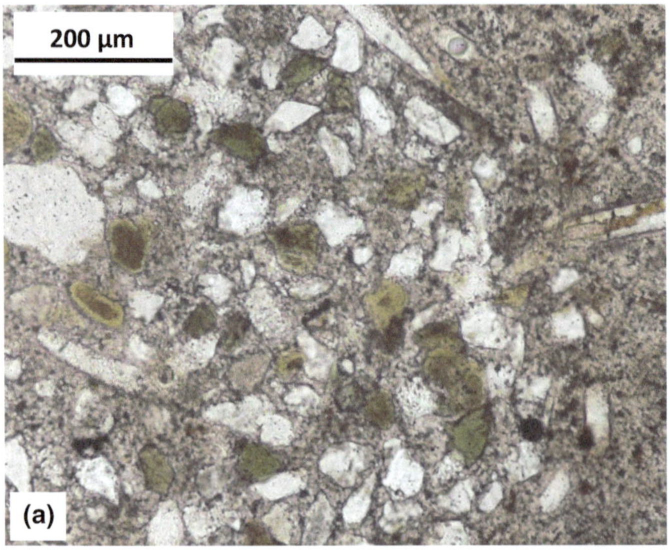

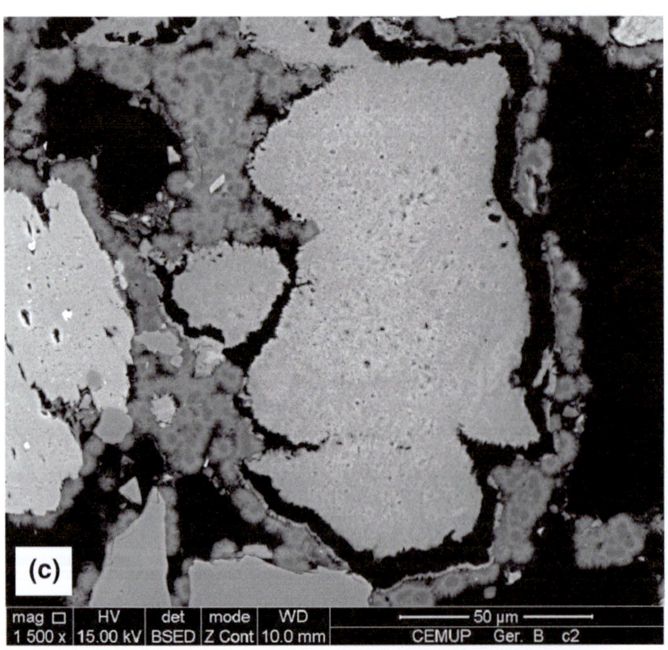

Sandstone (DE)

Poorly sorted marine opaline sandstone composed of sub-angular to sub-rounded detrital, coarse particles of monomineralic quartz, glauconite, as well as shell and polymineralic lithic fragments. The clasts are embedded in a muddy matrix comprising opaline silica, clay minerals and limonite. The glauconite pellets display a characteristic green to brownish-green colour (a, PPL), often with brown fringes due to oxidation of iron. The rock fragments mostly represent microcrystalline to cryptocrystalline chert. Siliceous cement is abundant and is composed of opal, appearing cloudy brownish from finely interspersed limonite (b, PPL). The sandstone is inhomogeneous, with patches where either glauconite or the opaline cement matrix is the main component. Prismatic sponge spicules commonly occur in the matrix. The thin sections were analysed by SEM (c, SEM-BSE): the massive light grey particle is quartz and the dark grey rims represent radiolarian debris in the matrix.

Reactivity potential attributed to:
Opaline silica, siliceous sponge spiculæ.

Sample of rock classified as potentially reactive in concrete structures and in laboratory tests (Germany).

Sandstone (FR)

Moderately-sorted, fine-grained sandstone composed of subrounded to angular clastic quartz, plagioclase, muscovite, tourmaline, chlorite, and zircon with interstitial organic matter and sericite. The sandstone is grain supported (a, XPL), in which the main component is monomineralic quartz particles of variable size. Larger clasts have oblong to concave-convex, interlobate boundaries (b, XPL). Sericite, chlorite and cryptocrystalline quartz of neogenic origin occur interstitially (c, XPL).

Reactivity potential attributed to:
Microcrystalline to cryptocrystalline quartz in clasts and matrix.

Aggregate particle in a sample of concrete which showed deleterious reactions in bridges and water structures (France).

Sandstone (NL)

Grain-supported coarse grained sandstone composed of sub-rounded monomineralic quartz, minor muscovite, biotite, K-feldspar, pyrite mostly altered to limonite, zircon and tourmaline (a, PPL). Grains are densely compacted, show interpenetrating boundaries (b, XPL) and are mostly overgrown with authigenic quartz in optical continuity. Original grain boundaries are only occasionally revealed by lining with clay minerals and/or percussion marks. There are microfractures crossing the clasts and also concentrated in the boundaries of the grains which are lined with iron (oxy-) hydroxides (limonite) (c, XPL). Intergranular spaces show a scarce matrix of clay minerals and some microcrystalline quartz, sericite and iron oxides.

Reactivity potential attributed to:
Overgrowths of authigenic quartz and microcrystalline siliceous rock cement.

Aggregate particle in a sample of concrete from a viaduct (The Netherlands).

Sandstone (NL)

Medium-sorted siliceous sandstone composed of monomineralic quartz and lithic fragments of quartzite, with interstitial microcrystalline quartz, sericite, and iron (oxy-) hydroxide (limonite) lining clast boundaries. The clasts are densely compacted and show variable grain size and shape (a, XPL). Quartz clasts are mainly (sub-) angular, with oblong to concave-convex contacts (b, XPL). Detrital mica flakes may be mildly kinked from compaction-deformation. Early-stage cracks are lined with insoluble minerals, limonite, and/or sericite (c, XPL) of diagenetic to very low-grade metamorphism.

Reactivity potential attributed to: *Interstitial microcrystalline quartz and compaction-deformed clasts.*

Aggregate particle in a sample of concrete from a viaduct (The Netherlands).

Sandstone (NL)

Well-sorted, grain-supported sandstone composed of angular particles mainly of mono-mineralic quartz and minor lithic fragments (quartzite, siltstone), coated by iron (oxy-) hydroxides and trace amounts of clay. The clasts have been densely compacted (a, XPL). Mud matrix and neogenic rock cement are both absent. Quartz grains display undulatory extinction and fluid inclusion trails sweeping across multiple grains (b, XPL). The clasts show serrate to interlobate contacts, locally interpenetrating (b, c, XPL), with incipient subgraining along grain peripheries.

Reactivity potential attributed to:
Microcrystalline to cryptocrystalline quartz in the rock fragments. Quartz with pressure-solution features.

Aggregate particle in a sample of concrete from a viaduct (The Netherlands).

Sandstone (NL)

Medium-sorted, medium-grained sandstone composed of monomineralic quartz with trace amounts of interstitial iron (oxy-) hydroxides (limonite) in the narrow intergranular spaces and in fractures (a, XPL). In this grain-supported very mature rock clastic particles are load-compacted and mildly deformed as shown by undulatory extinction. They are cemented by authigenic quartz in optical continuity (b, XPL). Original clast shape is completely obscured. Grain contacts are interlobate to intensely sutured (c, XPL).

Reactivity potential attributed to:
Strained quartz and interstitial microcrystalline quartz.

Aggregate particle in a sample of concrete from a viaduct (The Netherlands).

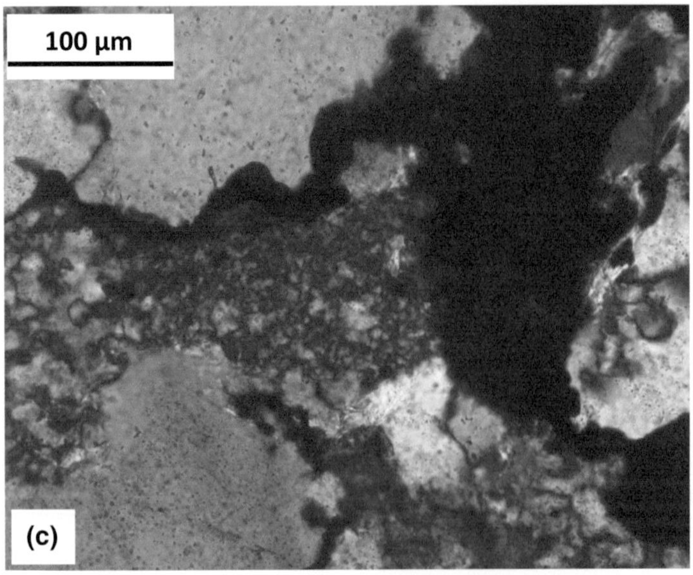

Sandstone (PT)

Poorly-sorted, coarse -grained, grain-supported sandstone composed of quartz and lithic fragments, stained by iron (oxy-) hydroxides (a, PPL). The clasts are subrounded to angular and elongate particles may locally define a poorly developed preferred orientation. Monomineralic quartz particles exhibit fluid inclusion trails (b, XPL), possibly strain lamellae, and some are internally cracked. The lithic fragments comprise microcrystalline to cryptocrystalline siliceous rocks. The rock is intensely stained with iron (oxy-) hydroxides (limonite) (c, XPL). Grain contacts are oblong and curved, but not observed to be serrated. Details are obscured by the strong limonite staining.

Reactivity potential attributed to:
Microcrystalline and cryptocrystalline quartz in clasts.

Aggregate particle in a sample of concrete from a dam (Portugal).

Sandstone (SE)

Grain-supported sandstone with particles of quartz, K-feldspar, plagioclase, minor muscovite, and traces of titanite and lithic fragments (mainly metamorphic rocks). This is a well-compacted clastic rock of moderately-sorted angular clasts (a, XPL), with locally concave-convex or serrate grain boundaries (b, XPL). Interstitial space is reduced and lined by iron (oxy-) hydroxides (limonite). There are clasts composed of microcrystalline to cryptocrystalline quartz. Clay minerals in the scarce matrix may be recrystallized to chlorite or sericite (c, XPL) (under very low-grade metamorphic conditions).

Reactivity potential attributed to:
Clasts of rock fragments with microcrystalline to cryptocrystalline quartz.

Aggregate particle in a sample of concrete from a dam (Sweden).

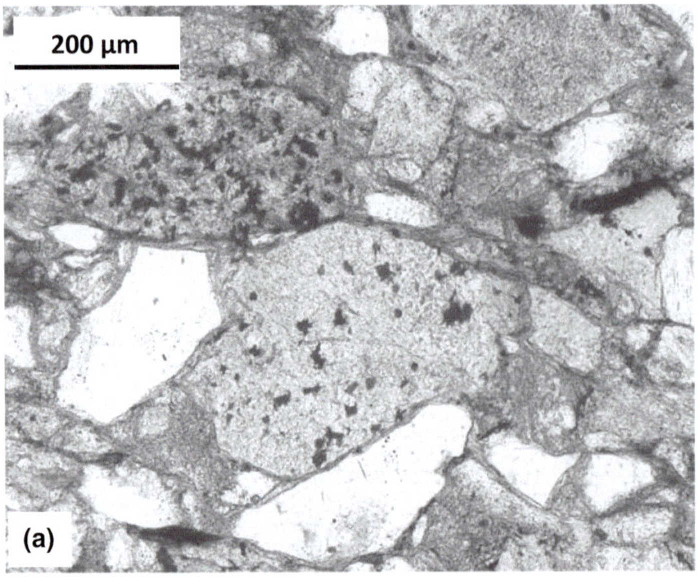

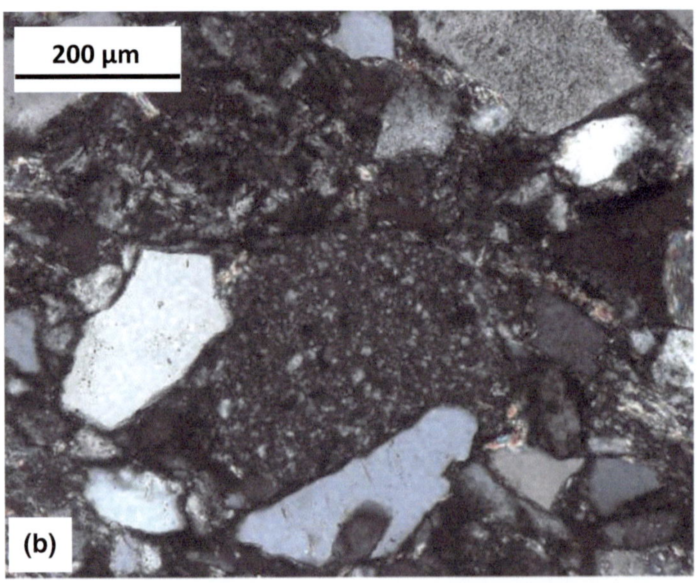

Sandstone (UK)

Poorly sorted clastic rock composed of angular to sub-rounded coarse monomineralic clasts of quartz, lithic fragments (e.g. fine-grained siltstone, chert), plagioclase and rarely K-feldspar. Minor sericite and traces of chlorite comprise the very fine-grained matrix (a, PPL). Rock cement is essentially composed of silica and carbonate. Lithic fragments may contain microcrystalline to cryptocrystalline quartz (b, XPL). Locally, matrix is scarce and the rock is predominantly grain-supported, with concave-convex grain boundaries (c, XPL). As large quartz clasts display uniform extinction, and feldspar twin lamellae and mica flakes are straight, deformation from load-compaction is minimal.

Reactivity potential attributed to:
Microcrystalline and cryptocrystalline quartz in the matrix and in lithic fragments.

Sample of rock classified as potentially reactive in laboratory tests and also in concrete structures such as dams, bridges and concrete road pavements (United Kingdom).

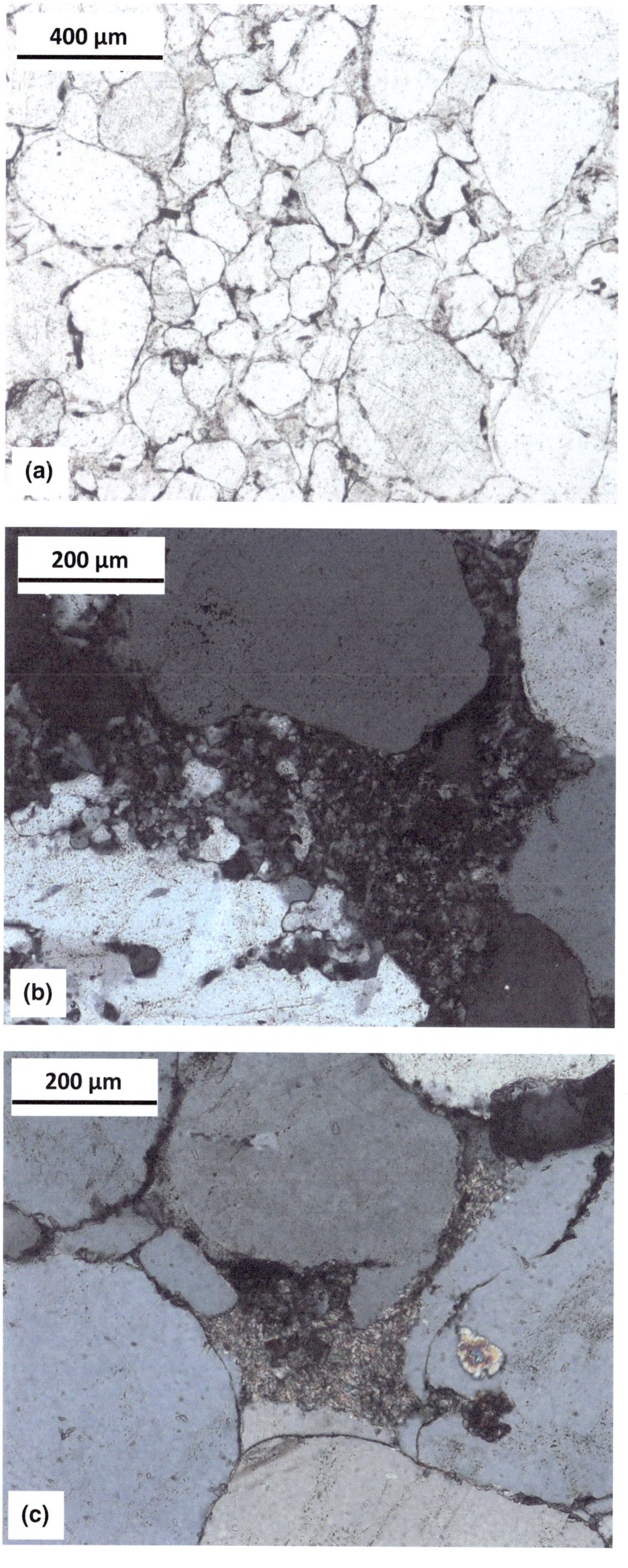

Sandstone (USA)

Sandstone composed of moderately-sorted, well-rounded clasts of quartz and lithic fragments (quartzite), with minor interspersed finer-grained particles of tourmaline, sericite, titanite and organic matter. Clastic particles are composed of monomineralic quartz (a, PPL). The lithic fragments present serrate to interlobate grain boundaries (b, XPL). Grain contacts are oblong to concave-convex, and the original periphery of clasts is lined with limonite. They are overgrown with authigenic quartz in optical continuity on clasts (c, XPL). Undulatory extinction is minimal or absent. The rock cement is siliceous and comprises microcrystalline to cryptocrystalline quartz. Minor sericite may occur locally, formed under diagenetic conditions (c, XPL).

Reactivity potential attributed to:
Microcrystalline to cryptocrystalline quartz of authigenic origin.

Aggregate particle in a sample of concrete prism (USA).

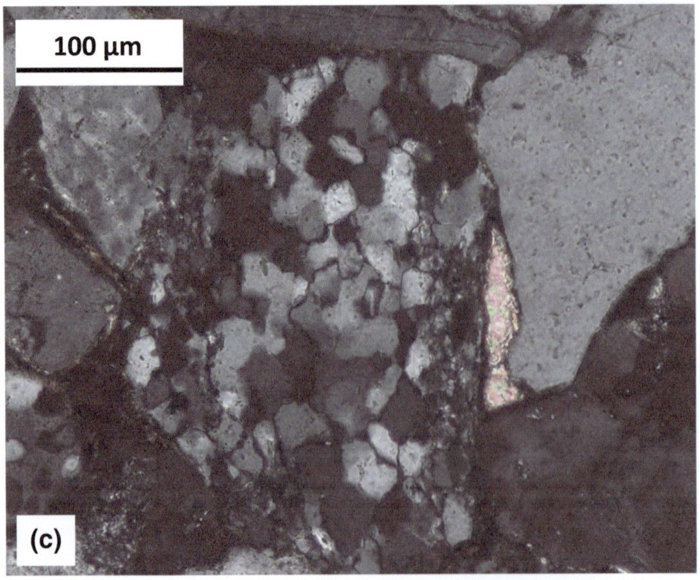

Sandstone (USA)

Poorly sorted, grain-supported arkosic sandstone composed of coarse angular clasts of quartz, plagioclase, K-feldspar, lithic fragments (e.g. quartz-phyllite, quartzite, chert, shale) (a, XPL) and organic matter. The fine-grained detrital matrix contains minor chloritised biotite. The sandstone is cemented by interstitial calcite (b, XPL), with minor cryptocrystalline silica. Feldspar clasts are cloudy from partial alteration to sericite or kaolinite. Lithic fragments of cryptocrystalline quartz (chert) and other rocks (e.g. quartzite) are abundant (c, XPL). The sandstone is well-compacted with many oblong and concave-convex contacts, pore spaces are filled with either detrital fine-grained matrix material or rock cement of neogenic origin.

Reactivity potential attributed to:
Abundant microcrystalline to cryptocrystalline quartz in chert rock fragments and in the rock cement.

Sample of rock classified as potentially reactive in laboratory tests (USA).

Sandstone (ZA)

Well-sorted sandstone, consisting almost entirely of monomineralic sub-angular quartz clasts with trace amounts of biotite, sericite, opaque minerals, titanite and zircon. This very mature, grain-supported sandstone shows a very high quartz content (a, XPL). The particles are cemented by secondary silica and some sericite and display undulatory extinction with serrate to interlobate boundaries. Inclusion trails cross multiple grains (b, XPL). The original shape of the clasts is obscured. Microcrystalline quartz along grain edges is deposited under compaction-diagenesis and very low-grade metamorphism (c, XPL).

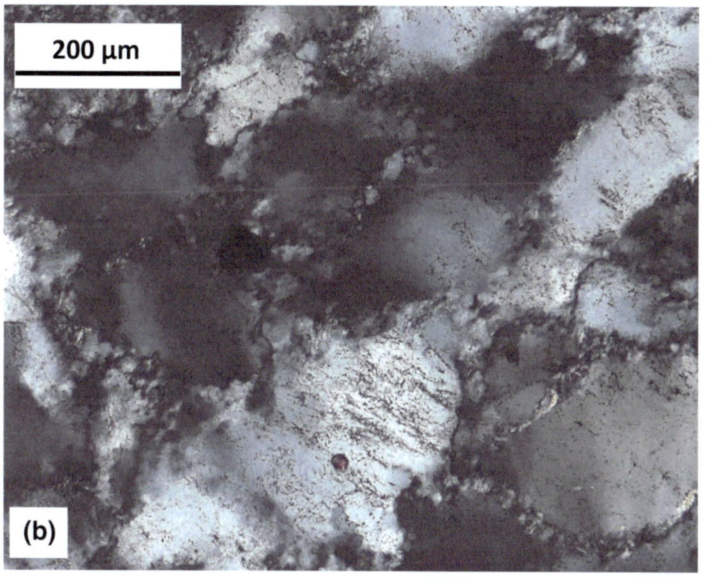

Reactivity potential attributed to:
Abundant microcrystalline to cryptocrystalline quartz along grain boundaries. Neogenic microcrystalline quartz rims.

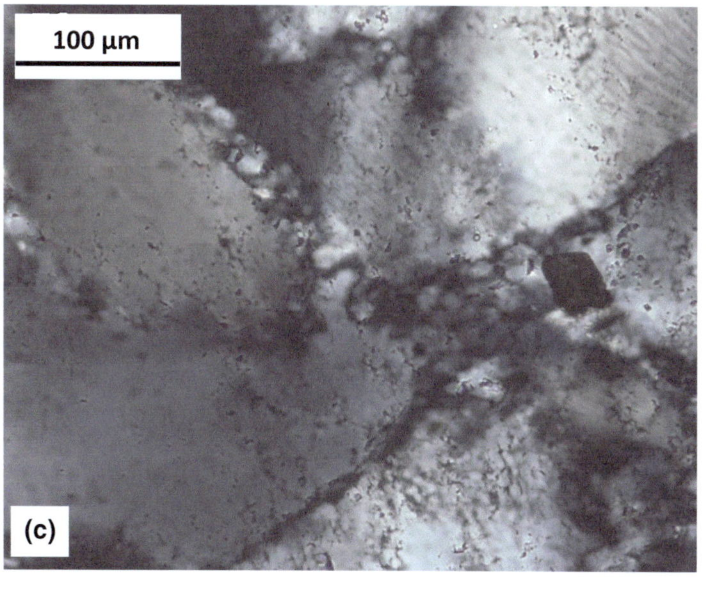

Aggregate particle in concrete from an airport apron (South Africa).

Sandstone (ZA)

Well-sorted sandstone, composed of medium to fine-grained quartz, with trace amounts of biotite, opaque minerals, titanite, zircon of detrital origin, locally associated with iron (oxy-) hydroxides ('limonite') (a, XPL). The originally well-rounded clasts of quartz show strong undulatory extinction and have serrate to interlobate grain boundaries with incipient subgraining (b, XPL). Some grains are stretched. The grains are firmly cemented by secondary silica. Pressure-solution phenomena and/or incipient tension-veining with elongate quartz are ubiquitous (c, XPL).

Reactivity potential attributed to:
Abundant strained, microcrystalline to cryptocrystalline quartz, subgraining.

Aggregate particle in concrete from an airport apron (South Africa).

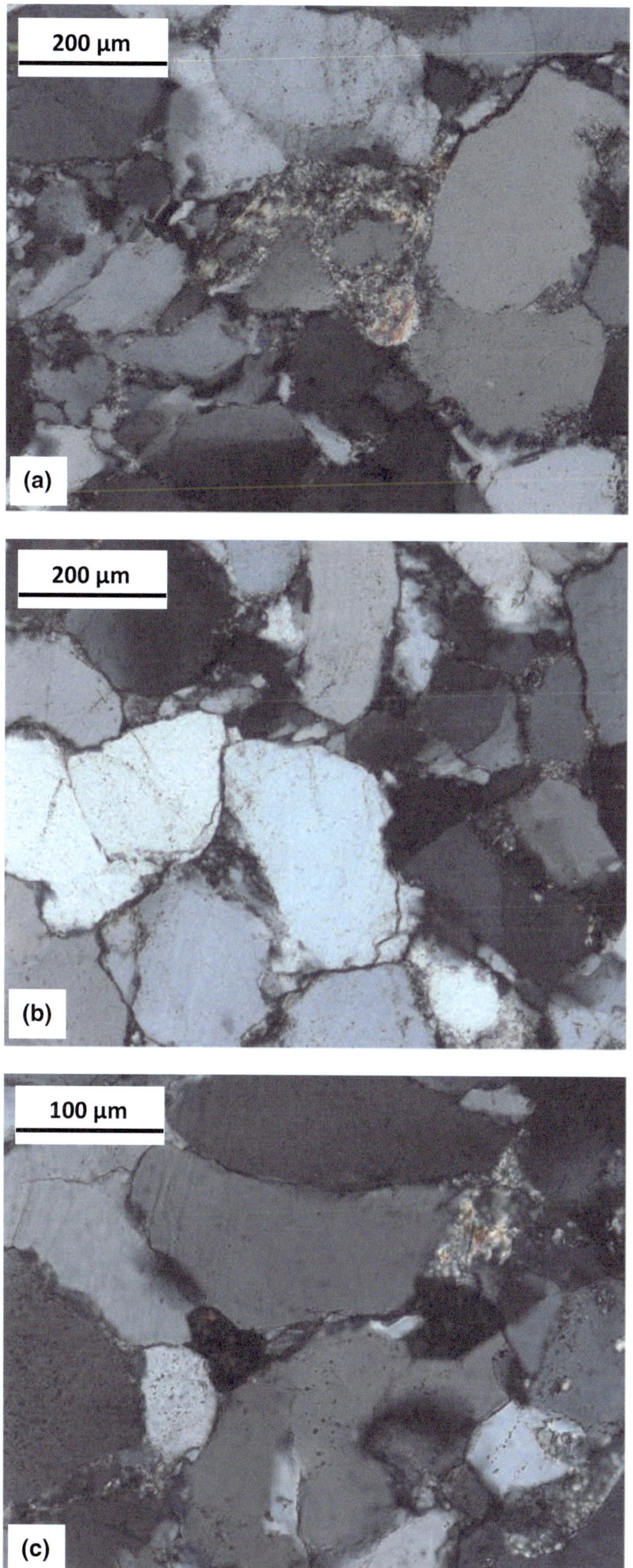

Sandstone (ZA)

Well-sorted coarse grained sandstone composed mainly of well-rounded quartz clasts of variable size with trace amounts of sericite, titanite, organic matter, iron (oxy-) hydroxides and lithic fragments (quartzite). Grain contacts are oblong to concave-convex (a, XPL). The rock is cemented by authigenic quartz deposited under diagenetic conditions, grown epitactically in optical continuity on clasts (b, XPL). Most of the quartz grains are monomineralic and show slight undulatory extinction. Sericite is found in the intergranular spaces (c, XPL).

Reactivity potential attributed to:
Epitactic overgrowth of authigenic quartz and neogenic siliceous rock cement.

Aggregate particle in concrete from an airport apron (South Africa).

Sandstone (ZA)

Poorly-sorted, fine-grained sandstone composed of main quartz, K-feldspar, plagioclase, muscovite, biotite, with minor opaque minerals and lithic fragments (chert and quartzite) (a, PPL). Quartz clasts range widely in size and are angular to sub-rounded, with elongate to convex grain contacts against quartz, straight against mica. Sub parallel orientation of detrital mica flakes defines a poorly developed diagenetic foliation (b, XPL). Biotite flakes are mostly altered to chlorite. K-feldspar is altered to kaolinite, plagioclase to sericite (c, XPL). Clasts are cemented by interstitial cryptocrystalline quartz of authigenic origin.

Reactivity potential attributed to:
Cryptocrystalline quartz of authigenic origin and also in chert clasts.

Aggregate particle in concrete from a bridge (South Africa).

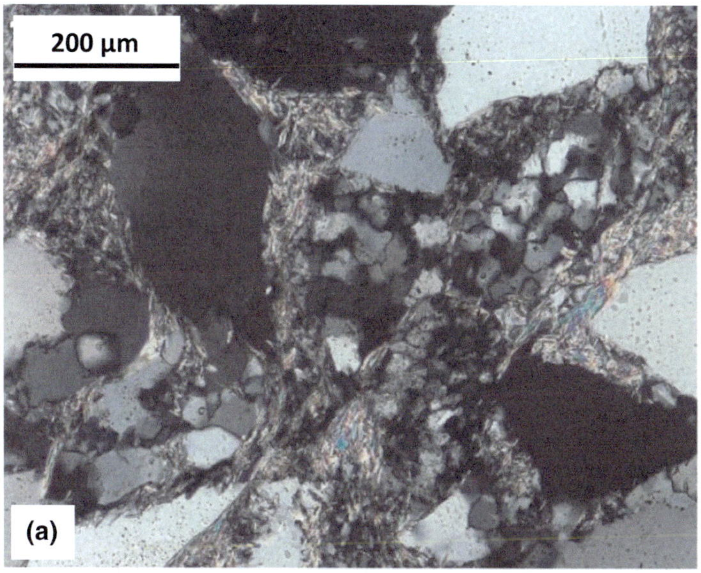

Sandstone (ZA)

Poorly-sorted rock composed of quartz, muscovite, and biotite, with trace amounts of opaque minerals, titanite and lithic fragments (quartzite) (a, XPL). The matrix contains abundant microcrystalline to cryptocrystalline quartz, chlorite, sericite and iron (oxy-) hydroxides (limonite). Large mica flakes may be kinked or bent from load-compaction (b, XPL), and lithic fragments display an internal polygonal fabric (c, XPL). The rock is cemented by neogenic microcrystalline quartz and sericite (b, XPL) formed under diagenetic to low-grade metamorphic conditions.

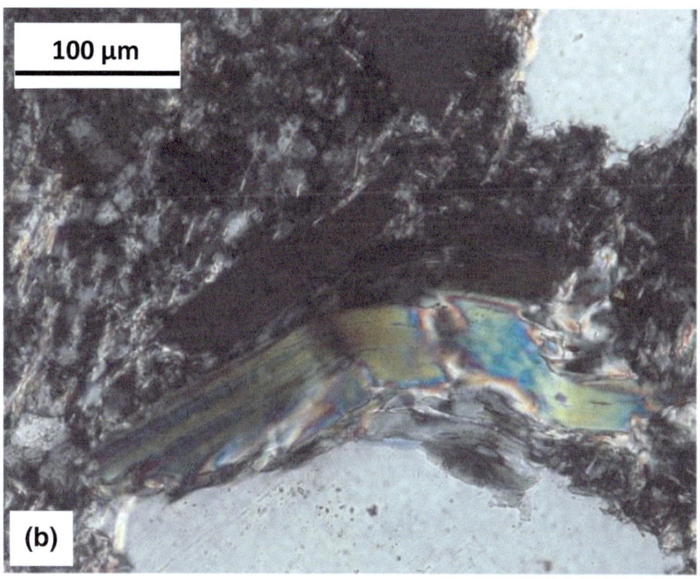

Reactivity potential attributed to:
Recrystallized and microcrystalline quartz.

Aggregate particle in concrete from a retaining wall (South Africa).

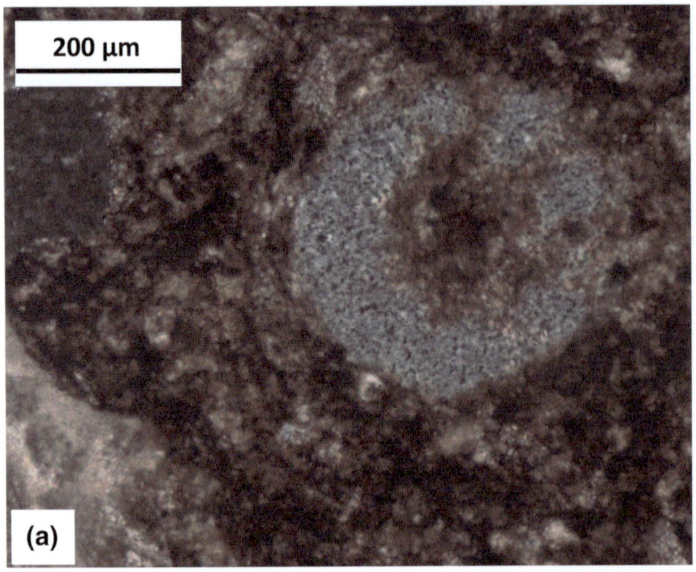

Siliceous limestone (BE)

Clastic argillaceous limestone composed of various forms of calcite, with trace amounts of cryptocrystalline quartz, clay minerals and organic matter (brownish) embedded in a micritic matrix (a, XPL). Bioclasts (fossil relics) of various types (e.g. crinoids, bryozoan, brachiopods, gastropods, bivalve, nautiloids) are abundant. They may be filled or partly replaced with sparry calcite (b, PPL), cryptocrystalline silica, or fibrous chalcedony of authigenic origin (c, XPL). Sparite and concretions of cryptocrystalline silica are observed to occur scarcely throughout the matrix. Compaction phenomena like pressure-solution and incipient stylolite formation are locally lined with insoluble clay minerals and/or limonite.

Reactivity potential attributed to:
Microcrystalline and cryptocrystalline quartz in the matrix and inside the bioclasts and some chalcedony.

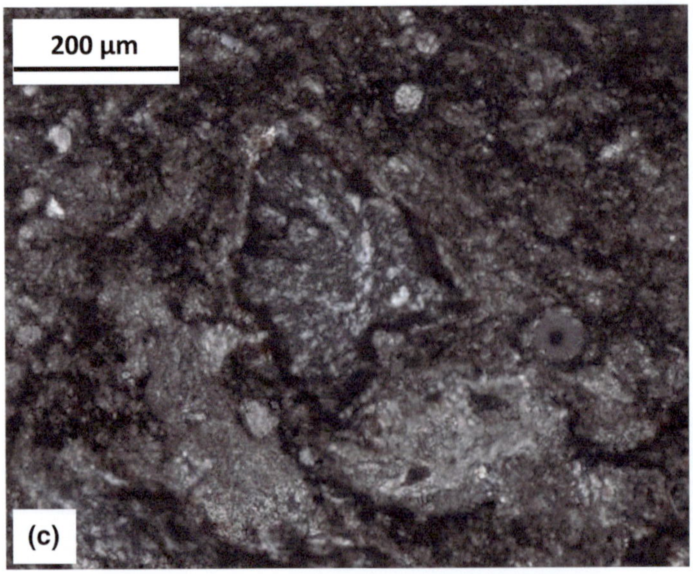

Sample of rock classified as potentially reactive in laboratory tests and also in concrete bridges and fluvial structures (Belgium).

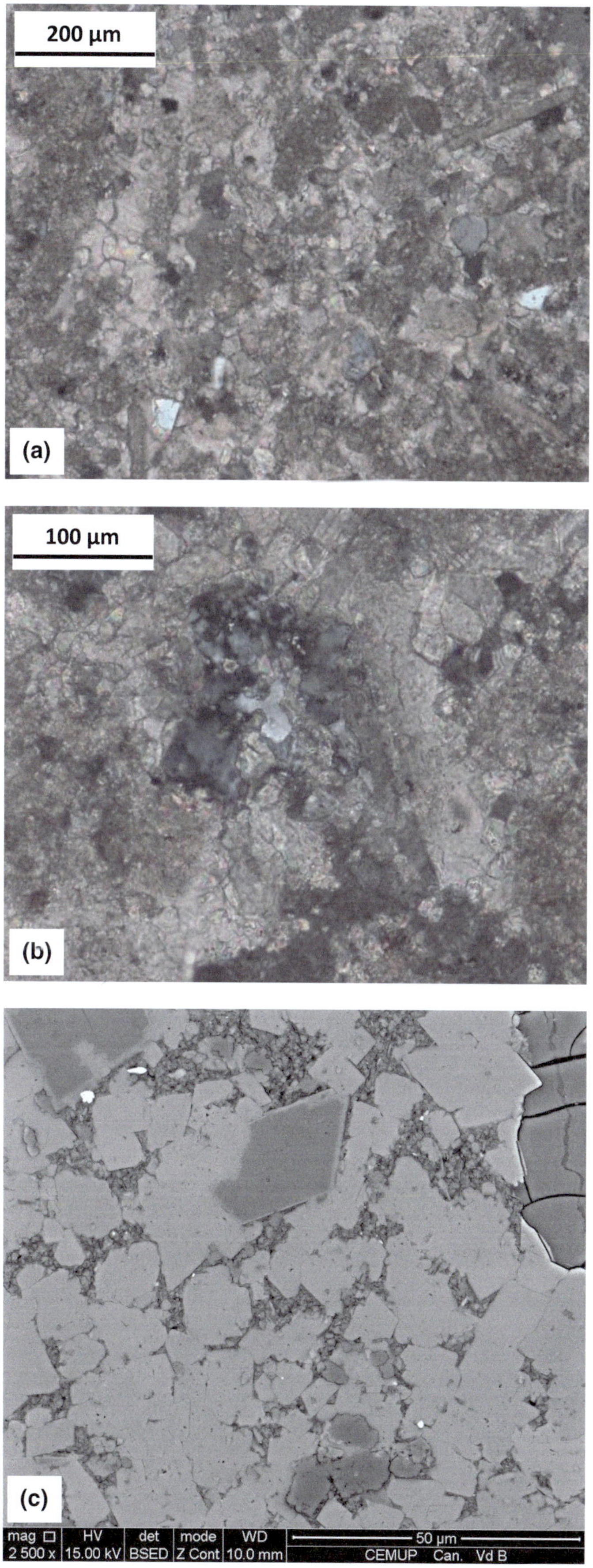

Siliceous limestone (CA)

Clastic siliceous dolomitic limestone composed of carbonate and siliceous bioclasts, with dolomite, monomineralic detrital quartz, with traces of interspersed clay minerals, locally stained reddish-brown by limonite. The matrix consists of brown-stained micritic calcite mud, locally recrystallized to sparite (a, XPL). Well-rounded clastic particles of detrital quartz are easily identified (a, XPL), as is chalcedony of neogenic origin filling voids (b, XPL). Backscatter imaging (c, SEM-BSE) confirms the porosity of the rock, the presence of sparite (light grey), dolomite (dark grey lozenges), interstitial clay minerals (yet darker grey particulate matter), and quartz (dark grey, irregular shapes).

Reactivity potential attributed to:
Microcrystalline and cryptocrystalline quartz.

Aggregate particle in a sample of concrete from a viaduct (Canada).

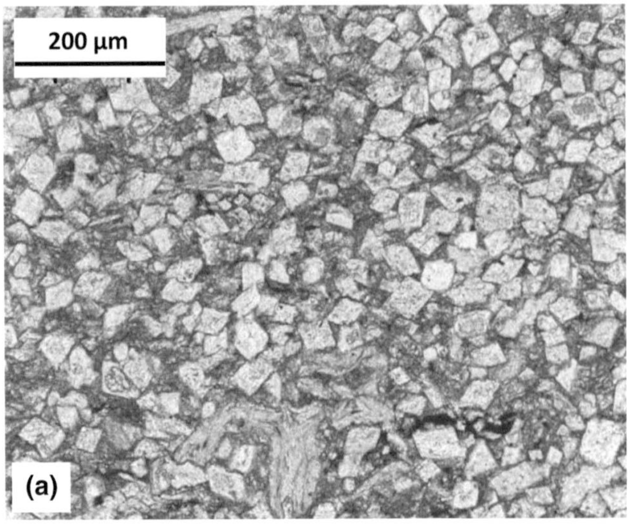

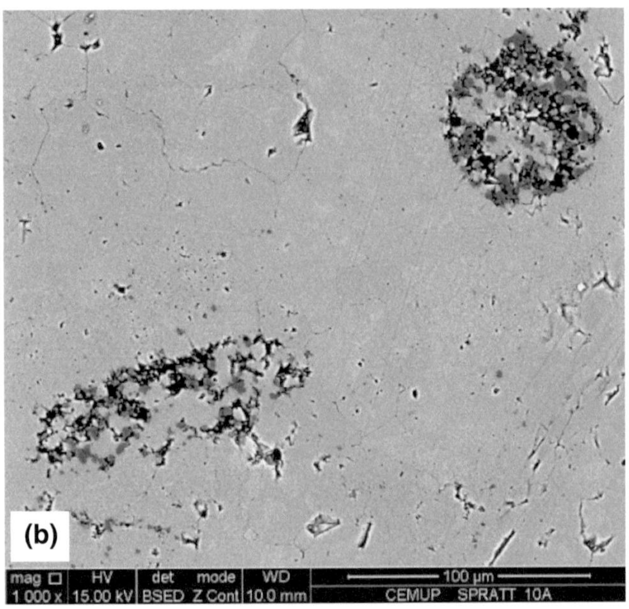

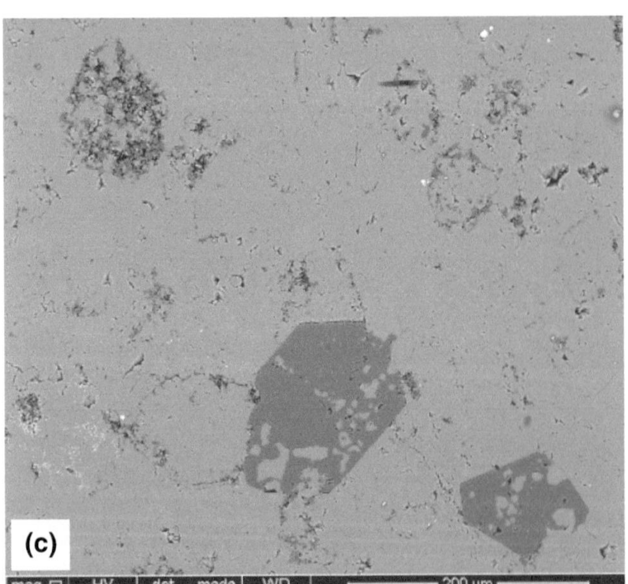

Siliceous limestone (CA)

Clastic, argillaceous dolomitic limestone composed of calcite, dolomite, carbonate bioclasts, and trace amounts of detrital clay minerals and chalcedony. Locally, euhedral, rhomb-shaped dolomite is predominant, embedded in a micritic matrix (carbonate mud) stained reddish-brown by clay and limonite (a, PPL). Elsewhere, microsparitic crystalline calcite is predominant and dolomite virtually absent, and very-fine grained silica and clay minerals occur in pockets (b, SEM-BSE). The rock has been recrystallized under diagenetic conditions, forming authigenic quartz with carbonate inclusions (c, SEM-BSE). The image also reveals the presence of siliceous matter in pockets, as well as finely disseminated throughout a carbonate matrix. Bulk whole-rock analysis confirms ~10wt% acid-insoluble content, of which ~9wt% is SiO_2 (according with literature), partly as quartz, partly as silicate in clay minerals.

Reactivity potential attributed to:
Microcrystalline to cryptocrystalline quartz (9% SiO_2) (according to literature) not discernible in optical thin section petrography.

Aggregate particle in a sample of concrete prism (Canada).

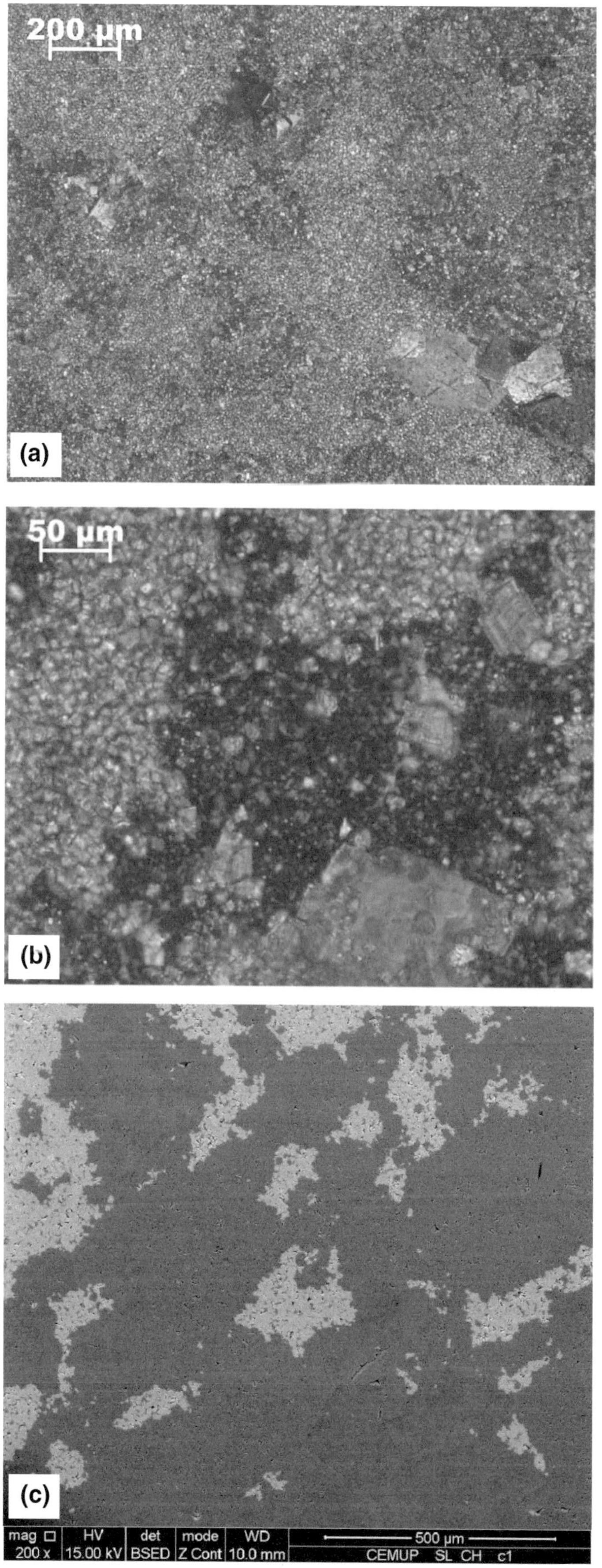

Siliceous limestone (CN)

Siliceous limestone composed of crystalline calcite, minor dolomite, and interspersed cryptocrystalline quartz. Some domains of the rock consist of very-fine grained calcite in which euhedral, lozenge-shaped crystals of dolomite are easily identified (a, XPL). The cryptocrystalline quartz occurs partly inconspicuously dispersed throughout the carbonate, partly also as clusters (b, XPL). Observation using electron microscope (c, SEM-BSE) confirms clustering of silica (dark grey) in carbonate (light grey).

Reactivity potential attributed to:
Microcrystalline to cryptocrystalline quartz disseminated throughout the carbonate matrix.

Rock classified as potentially reactive in laboratory expansion tests (China).

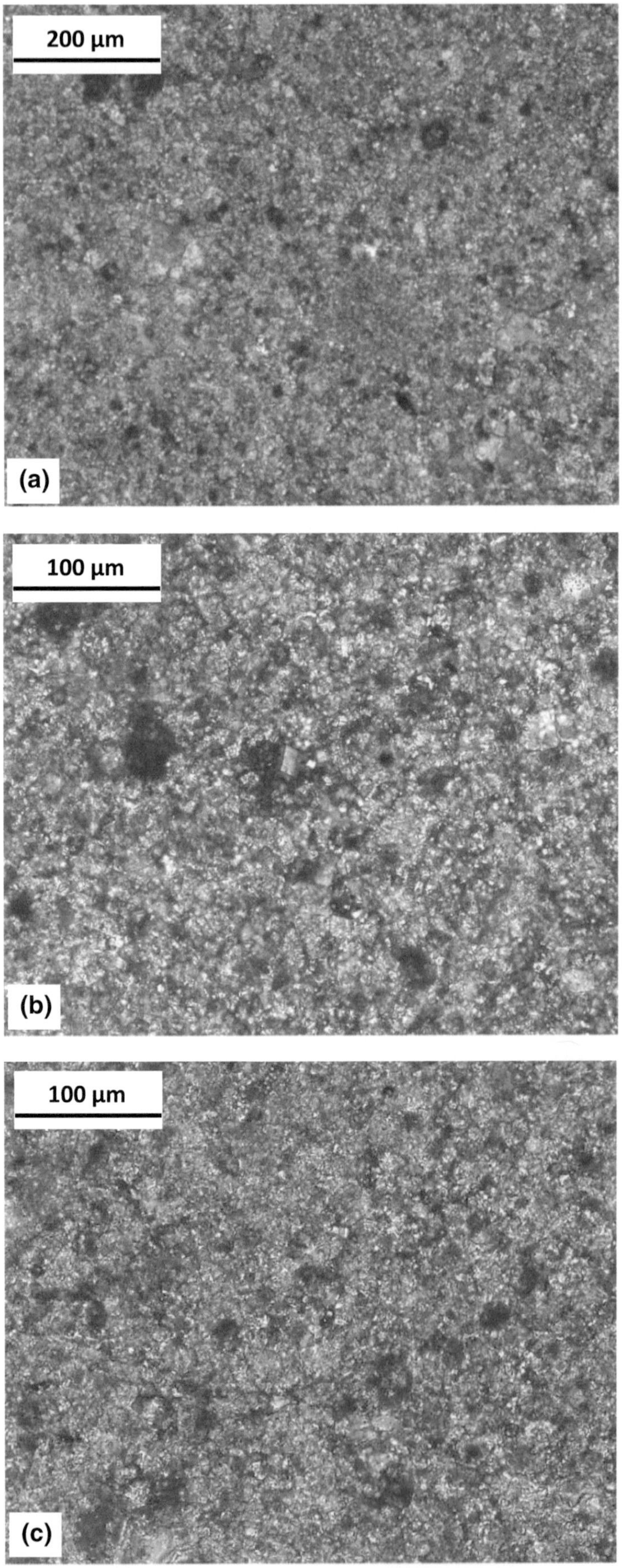

Siliceous limestone (DE)

Bioclastic, very-fine grained siliceous limestone composed of calcite with minor dolomite, trace amounts of cryptocrystalline quartz and finely interspersed reddish-brown iron (oxy-) hydroxides (limonite) (a, XPL). The rock contains bioclastic fragments and ooliths embedded in a fine-grained micritic matrix, locally with limonite concretions and sparry calcite (b, XPL). Microcrystalline to cryptocrystalline silica clasts occur dispersed throughout the matrix (b, XPL). Sparite may also occur within bioclasts. In the high-birefringent carbonate matrix, quartz is barely distinguishable (c, XPL).

Reactivity potential attributed to:
Microcrystalline and cryptocrystalline quartz dispersed throughout the carbonate matrix.

Sample of rock classified as potentially reactive in laboratory tests and also in concrete pavements (Germany).

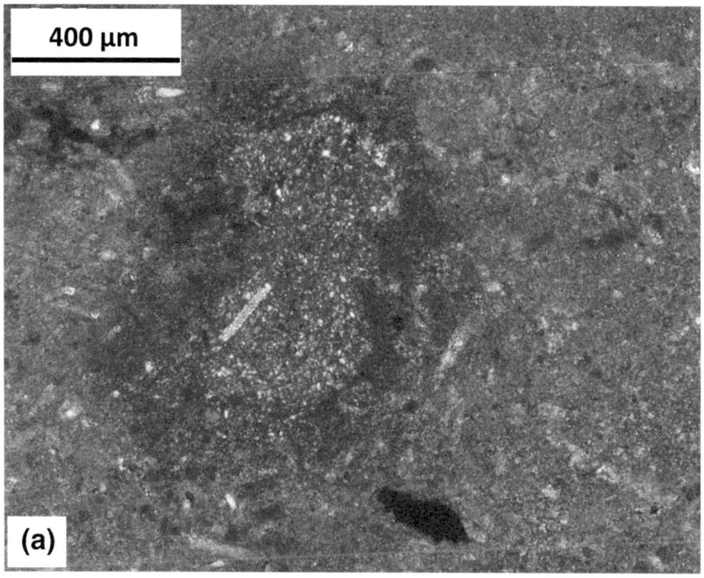

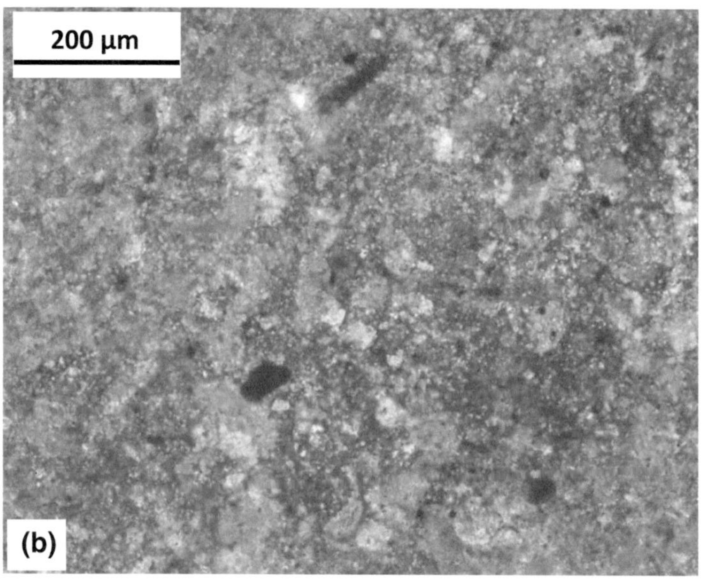

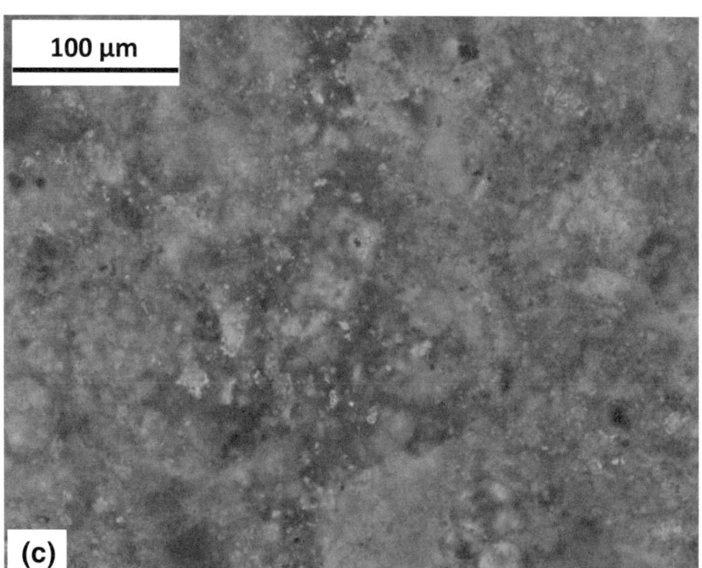

Siliceous limestone (FR)

Siliceous limestone composed of micritic calcite mud and sparite, shell fragments and trace amounts of cryptocrystalline silica in the cement and replacing or filling bioclasts (a, XPL). This very fine-grained micritic limestone contains abundant carbonate bioclasts, as well as siliceous sponge spiculae, inconspicuously dispersed throughout the high-birefringent micrite matrix (b, XPL). Some bioclasts have been partly replaced or filled with cryptocrystalline silica of neogenic origin; others have been replaced with sparite (b, XPL). Reddish-brown stains are caused by limonite. Sparry calcite (crystalline) occurs in the cryptocrystalline quartz-rich domains (c, XPL).

Reactivity potential attributed to:
Microcrystalline and cryptocrystalline quartz.

Aggregate particle of sand proved to give alkali-silica reaction in concrete bridges and water structures (France).

Siliceous limestone (FR)

Siliceous limestone composed of micritic carbonate mud and bioclasts (mostly calcite). Voids are filled with fibrous chalcedony, and locally stained by iron (oxy-) hydroxides (limonite) and possibly detrital clay minerals (a, XPL). Scarce lozenge-shaped dolomite crystals occur as well. The abundant hollow fossil relics are filled with chalcedony and/or cryptocrystalline quartz (b and c, XPL). They may also contain opaline silica.

Reactivity potential attributed to:
Cryptocrystalline quartz, opaline silica and chalcedony.

Lithology known to develop deleterious ASR in bridges and water structures (France).

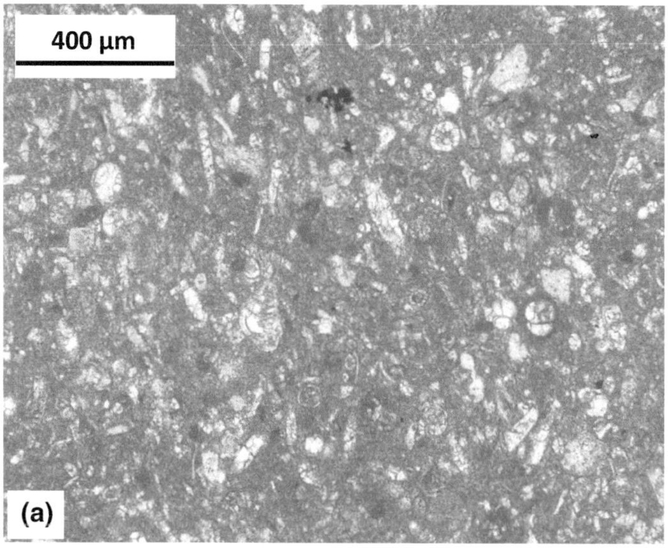

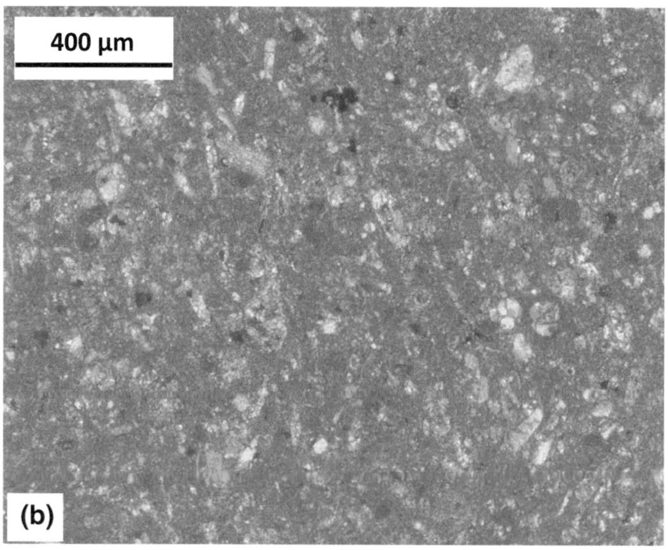

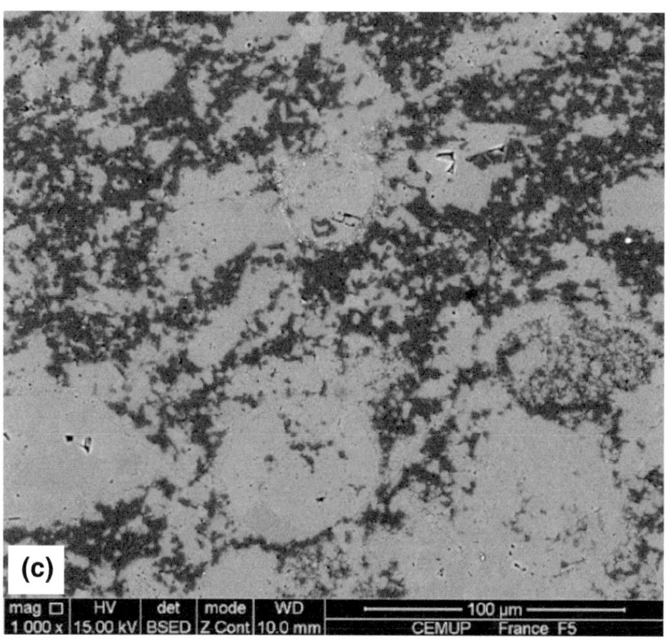

Siliceous limestone (FR)

Very fine to fine-grained clastic limestone composed of abundant bioclasts, sparry calcite and detrital quartz grains embedded in micrite carbonate cement coloured grey by finely dispersed clay minerals (a, PPL). Bioclasts comprise foraminifera, shell fragments and others, which have been partly replaced by crystalline sparite (b, XPL). Silt-fraction sized detrital particles of quartz occur finely dispersed throughout the micritic matrix (b, XPL). Backscatter imaging (c, SEM-BSE) confirms the presence of sparitic calcite (light grey) in bioclasts, and cryptocrystalline silica (dark grey).

Reactivity potential attributed to:
Microcrystalline to cryptocrystalline quartz.

Particles in sand aggregate proved to give alkali-silica reaction in concrete bridges and water structures (France).

Metamorphic Rocks

4

Isabel Fernandes, Maria dos Anjos Ribeiro, Maarten A.T.M. Broekmans, and Ian Sims

Metamorphic rocks form from prograde or retrograde recrystallization of igneous, sedimentary or metamorphic precursor rocks, variously changing mineral content, structure, and texture/fabric. Metamorphic rocks may be formed from shallow depth near the Earth's surface to great depth down in the mantle. Weathering and alteration resulting from (near) surface conditions are by definition excluded. When bulk whole-rock chemistry changes during metamorphism, then the process is called metasomatism instead. Classification of metamorphic rocks is complicated and is based on:

(a) mineral (modal) content,
(b) microstructure,
(c) texture/fabric,
(d) inferred origin (protolith),
(e) metamorphic context, i.e. pressure, temperature, deformation, presence of fluids,
(f) bulk whole-rock chemical composition.

This atlas uses the IUGS Subcommission system described in Fettes and Desmons (2007) for the classification of metamorphic rock types. This classification is based on the division into three major groups on the basis of their structure (as seen in hand specimen)—schist, gneiss and granofels. However, other designations are here applied, based on the same document, defined in respect of specific factors: e.g. quartzite (more than 75 % of quartz), mylonite, fault breccia, pseudotachylite and cataclasite when the context and/or genetic process of the rock are known and are to be emphasised.

According to this IUGS system, a given metamorphic rock may actually have different correct names, either referring to its protolith, to a systematic name, or to a generic name. Which name the petrographer prefers depends on available information on geological setting and on which aspect is considered to be the most relevant for the actual context. Thus, one and the same rock sample may be classified as 'meta-basalt' if the protolith can be established, or as 'hornblende-plagioclase gneiss' if it has a banded structure (at hand specimen scale), or instead generically as 'amphibolite' if geological setting and macro-structure are unknown or uncertain. As another example, a 'meta-limestone' in a regionally metamorphosed sedimentary sequence may alternatively be named as 'calc-silicate granofels' if it has a non-directional fabric, or generically as 'marble' in case its origin remains undisclosed.

The rock nomenclature as used in this atlas reflects content in known alkali-reactive mineral species and their spatial distribution (i.e. microstructure, texture/fabric), which are known to determine to a large extent the alkali-reactivity potential. The fine grain size of microcrystalline quartz (<100 μm) presents a large potential reaction surface area, but that of cryptocrystalline quartz (<10 μm) is even larger. Deformation of quartz under ductile (as in mylonite) or brittle (as in cataclasite or fault breccia) conditions introduces stress in the crystal lattice, visible as undulatory extinction, evolving into subgraining for larger amounts of strain.

Each of the main groups comprises different rock types. The groups may seem to have an uneven number of samples but the actual number of samples presented depends on the potential reactivity related to that group and, on the other hand, on the abundance of each rock type in what concerns its application as concrete aggregates. The subdivision of the

I. Fernandes (✉)
ICT, Institute of Earth Sciences and Department of Geology, Faculty of Sciences, University of Lisbon, Lisbon, Portugal
e-mail: mifernandes@fc.ul.pt

M.A. Ribeiro
ICT, Institute of Earth Sciences and DGAOT, Faculty of Sciences, University of Porto, Porto, Portugal
e-mail: maribeir@fc.up.pt

M.A.T.M. Broekmans
Department of Mineral Resources, Geological Survey of Norway - NGU, Trondheim, Norway
e-mail: maarten.broekmans@ngu.no

I. Sims
RSK Environment Ltd, Hemel Hempstead, UK
e-mail: ian@simsdoc.com; isims@rsk.co.uk

I. Fernandes et al. (eds.), *Petrographic Atlas: Characterisation of Aggregates Regarding Potential Reactivity to Alkalis*, DOI 10.1007/978-94-017-7383-6_4

Table 4.1 Main types of metamorphic rocks in this atlas, based on the classification of Fettes and Desmons (2007)

Granofels	Rocks with granoblastic texture: hornfels, granulites and calc-silicate rocks
Gneisses	Rocks with characteristic banded 'gneissose' structure in which there are segregations of light colour minerals (usually quartz and feldspar) and dark colour minerals (biotite and amphibole). This group usually presents deformed, stretched quartz, with subgrains and recrystallization
Slates	Very fine-grained rocks with well-developed regular fissility along planes (slaty cleavage)
Schists	Rocks displaying on the hand-specimen scale a pervasive foliation, well-developed cleavage or schistosity defined by the preferred orientation of abundant inequant mineral grains with foliation, such as the phyllites (fine-grained rocks) and schists (medium- to coarse-grained varieties). In most of these rocks phyllosilicates define the preferred orientation of the rock exhibiting a lepidoblastic texture
Tectonites	Rocks in which the main characteristic is a structure marked by tectonic deformation, such as fault breccia, cataclasite and pseudotachylite (brittle regime) and mylonite (ductile regime)

groups according to the aggregates that were studied is presented in Table 4.1.

Groups represented by a higher number of samples are gneisses and tectonites, with different extents of deformation as well as different deformation regimes. Besides microcrystalline and cryptocrystalline quartz, deformed quartz seems to be the main cause of alkali-reactivity in metamorphic rocks. According to Passchier and Trouw (2005) (Fig. 4.1), deformation is exhibited by undulatory extinction, strain lamellae, sub-grain formation, and dynamic recrystallization.

Recrystallization of deformed quartz reduces grain size and lattice strain by the formation of subgrains, in which the lattice is rotated with respect to the parent and separated by a high-angle grain boundary. Whereas complete recrystallization is able to remove undulatory extinction, it does not necessarily remove the risk of alkali-reactivity (Broekmans 2002).

Thin section petrography of metamorphic rocks thus requires detailed assessment not only of mineral content, but also of (micro-) structure, texture and fabric, with particular emphasis on quartz, including its grain size, the presence of undulatory extinction, low-angle grain boundaries or subgrains, and deformation lamellae, as well as the morphology of boundaries between adjacent quartz grains (convex, serrate, interlobate) and the overall mineral fabric.

References

Broekmans, MATM (2002): The alkali-silica reaction: mineralogical and geochemical aspects of some Dutch concretes and Norwegian mylonites. PhD Thesis, University of Utrecht. Geologica Ultraiectina (217): pp 144.

Fettes, D, Desmons, J (editors) (2007): Metamorphic Rocks: A Classification and Glossary of Terms. Recommendations of the International Union of Geological Sciences Subcommission on the Systematics of Metamorphic Rocks. British Geological Survey. ISBN-13: 9780521868105.

Passchier, CW, Trouw, RAJ (2005): Microtectonics. 2nd edition. Springer Verlag: pp 354.

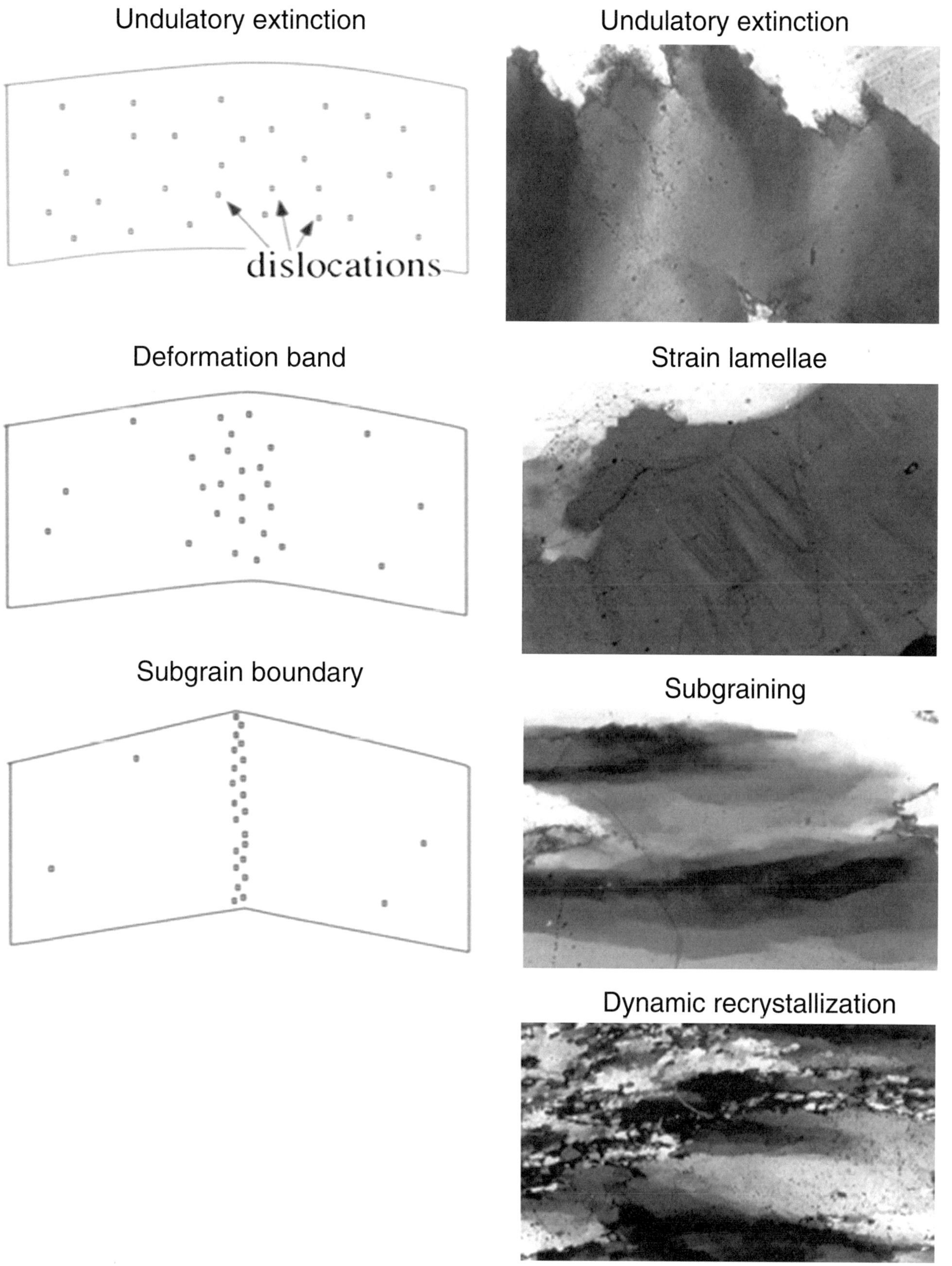

Fig. 4.1 Schematic representation and illustrative images of the different steps of progressive deformation and dynamic recrystallization of quartz (Adapted from Passchier and Trouw 2005)

Amphibolite (PT)

Meta -igneous rock composed of amphibole (hornblende), plagioclase, quartz, epidote, calcite and titanite. The rock is crossed by veins comprising quartz, amphibole, chlorite and calcite. This amphibolite exhibits foliation marked by the amphibole prismatic blasts (greenish) defining a nematoblastic texture (a, PPL). Quartz, although scarce, is mainly microcrystalline and occurs in aggregates of cryst als with indented boundaries. In addition, there are abundant shear zone with microcrystalline and cryptocrystalline quartz (b, XPL). There are numerous thin veins composed of comb -like quartz associated with amphibole and also cryptocrystalline quartz (c, XPL).

Reactivity potential attributed to:
Microcrystalline quartz in the interstitial spaces. Cryptocrystalline quartz in the boundaries of the veinlets and associated with shear zones (faults)..

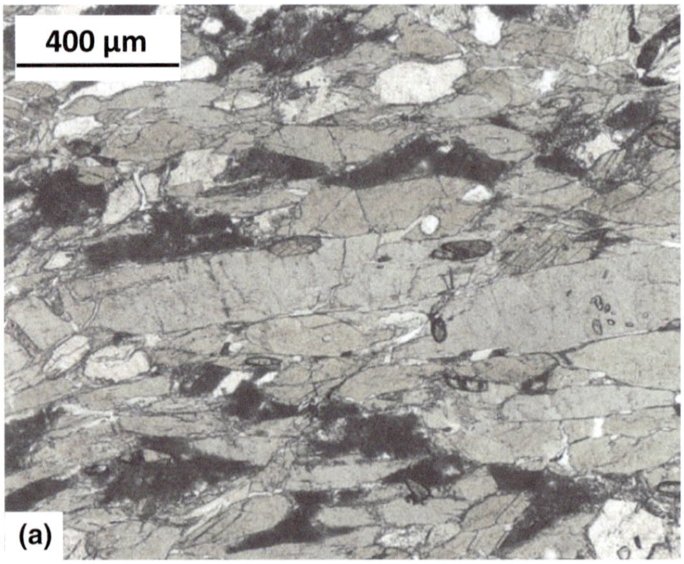

Sample of rock found to be potentially reactive in concrete prism tests (Portugal).

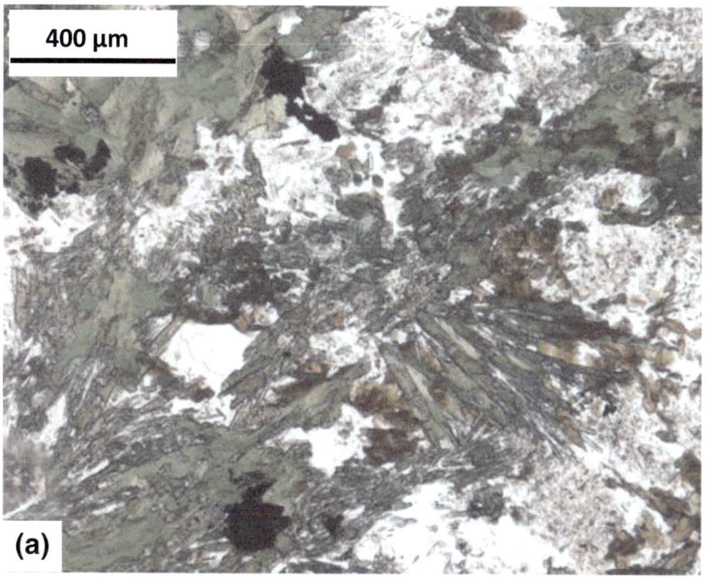

Calc-silicate rock (FI)

Calc-silicate rock composed of quartz, plagioclase, tremolite-actinolite, opaque minerals, biotite, calcite and epidote.
The rock exhibits granoblastic texture with domains where fibrous crystals of amphibole (tremolite-actinolite) form agglomerations of radial texture (a, PPL). Microcrystalline quartz is observed in the boundaries of the larger crystals and sometimes moulding the porphyroblasts of plagioclase (b, XPL). Larger crystals of quartz exhibit straight to indented boundaries and recrystallization. Locally microcrystalline quartz forms mosaics of polygonal shapes (c, XPL).

Reactivity potential attributed to:
Microcrystalline quartz in the boundaries of larger blasts. Quartz shows indented boundaries.

Aggregate particle in a sample of concrete from a paper mill (Finland).

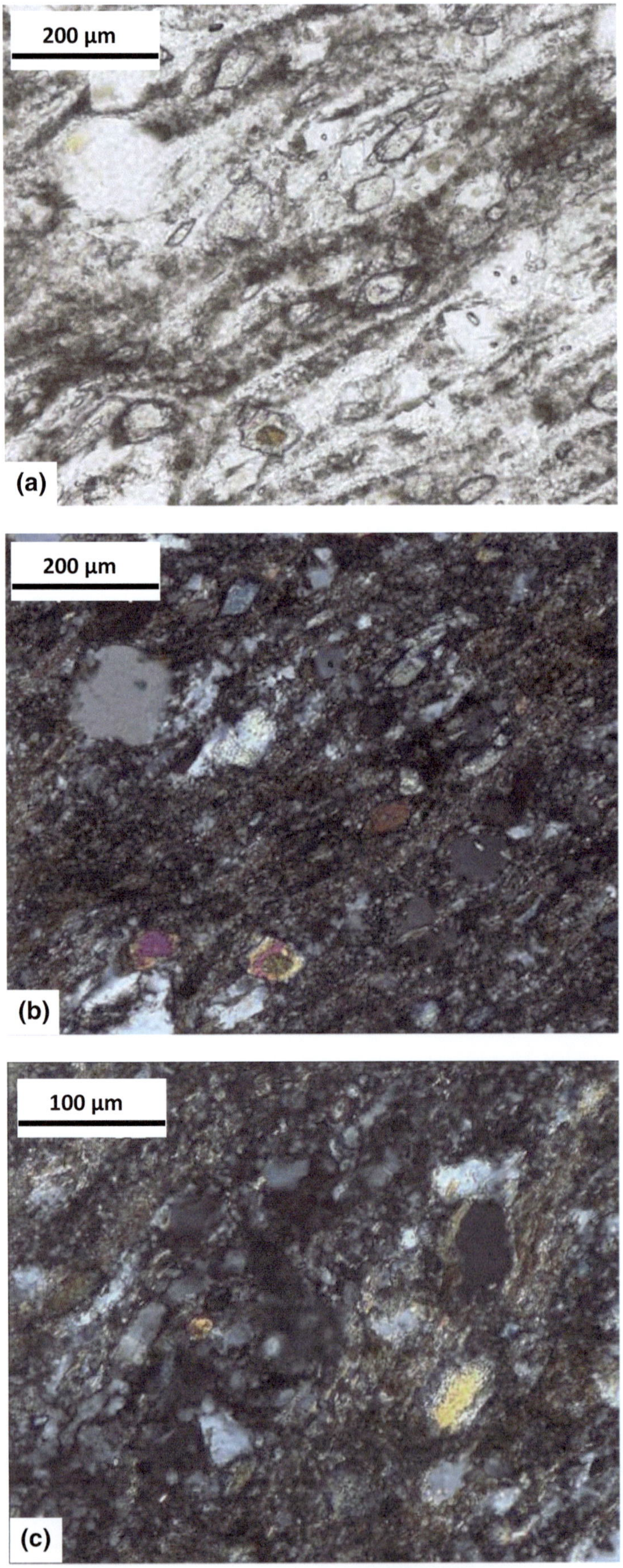

Calc-silicate rock (NO)

Metasedimentary calc-silicate rock composed of quartz, biotite, chlorite, epidote and iron hydroxides. This rock of low metamorphic grade exhibits heterogranoblastic texture and abundant matrix. Biotite defines a preferred orientation and occurs as clots which might have generated by the recrystallization of clay minerals in the matrix of the original rock (a, PPL). Coarse quartz grains are occasional and the main components occur as very fine-grained (b, XPL). There are also agglomerations of microcrystalline quartz in rock fragments inherited from the sedimentary clasts. Microcrystalline to cryptocrystalline quartz is the main component of the rock (c, XPL).

Reactivity potential attributed to:
Microcrystalline to cryptocrystalline quartz.

Aggregate particle in a sample of concrete prism. Deleterious ASR with this aggregate was observed in the concrete of a bridge (Norway).

Calc-silicate rock (ZA)

Fine-grained granofels, composed of mainly quartz, epidote, calcic clinopyroxene altered to calcic amphibole, with minor plagioclase, sericite, and accessory calcite. The constituent minerals are arranged in a random, non-directional granoblastic texture (a, XPL) with local polygonal grain contacts. The abundant microcrystalline quartz occurs in variable size with concave-convex to serrate grain boundaries (b, XPL). Epidote occurs abundantly in polycrystalline clusters throughout the rock. Amphibole crystals contain relics of its pyroxene precursor, sericite and calcite may be of later origin, possibly retrograde from plagioclase (c, XPL).

Reactivity potential attributed to:
Microcrystalline quartz, cryptocrystalline quartz interspersed in amphibole as by-product from pyroxene alteration.

Aggregate sample from a formation classified as potentially reactive (South Africa).

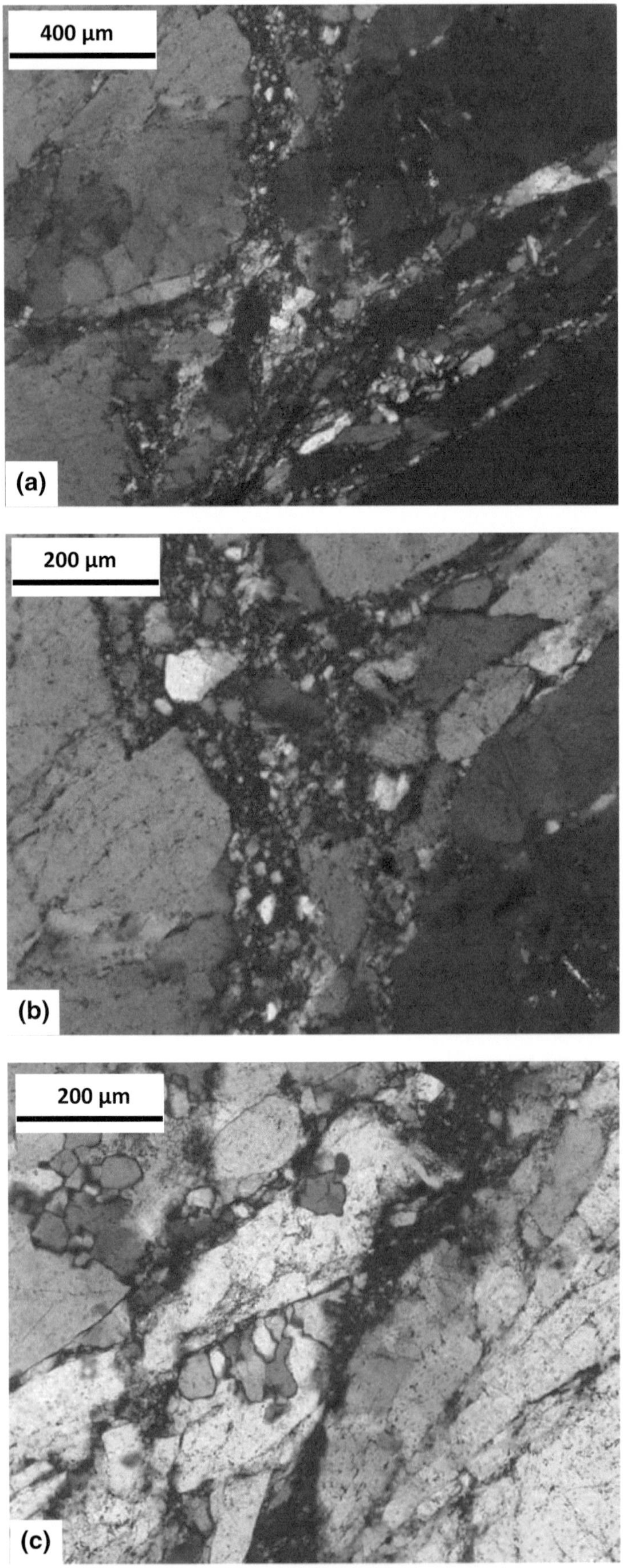

Cataclasite (AU)

Cataclasite composed of quartz and feldspar fragments, with minor sericite and clay minerals of secondary origin, locally lined by limonite. Coarse and fine fragments from fault breccia owing to brittle tectonic deformation are consolidated by neogenic quartz, sericite and clay minerals (a, XPL). Intense subgraining occurs in alignments of ductile deformation with lateral branches (a and b, XPL). Quartz recrystallization is reflected in bulging along the boundaries growing perpendicularly to the previous alignments (c, XPL).

Reactivity potential attributed to:
Microcrystalline quartz fragments from brittle deformation, undulatory extinction and subgraining from ductile deformation, cryptocrystalline quartz cement.

Aggregate particle in a sample of concrete from a hydraulic structure (Australia).

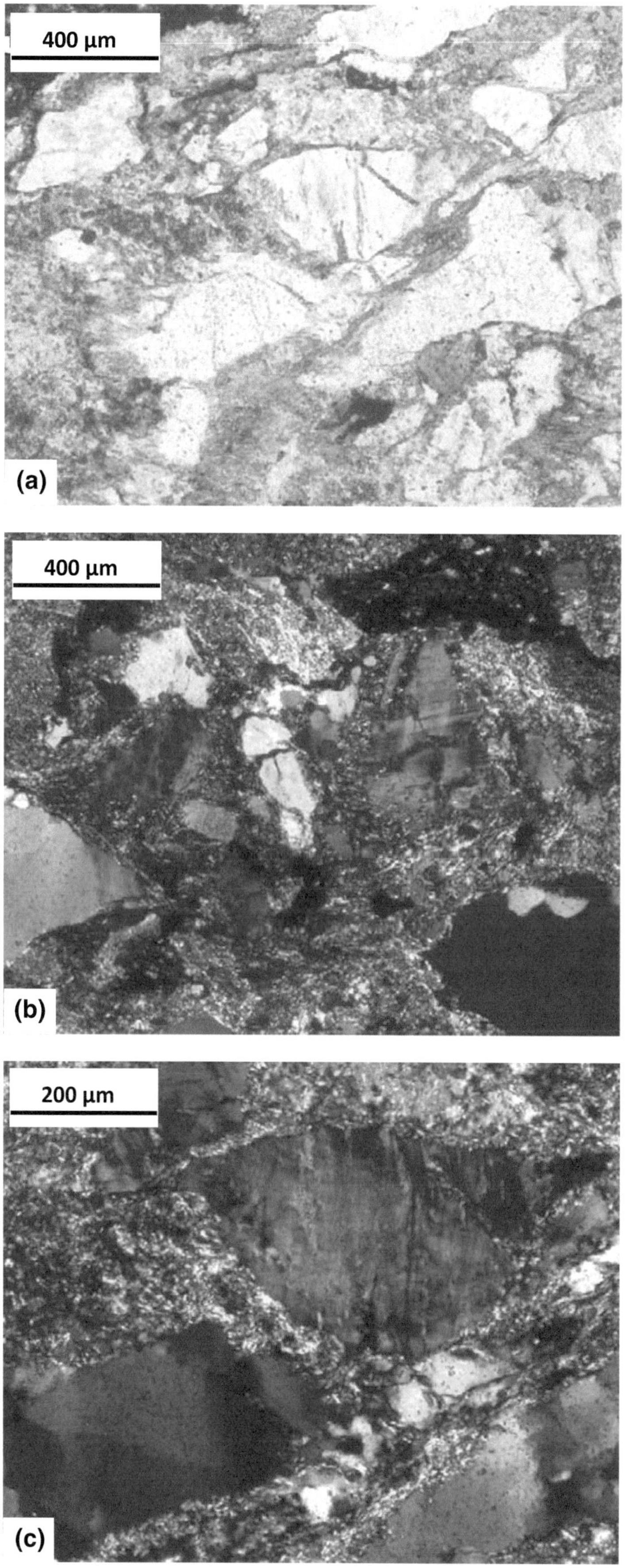

Cataclasite (AU)

Coarse-grained metamorphic rock composed of quartz, plagioclase, K-feldspar, with minor sericite, and trace amounts of epidote, chlorite, clay minerals, and opaque minerals. The feldspars appear cloudy from partial alteration to sericite and clay minerals (a, PPL). Epidote and chlorite occur in the matrix, often forming fine-grained polycrystalline clusters. Quartz fragments display strong undulatory extinction, feldspar fragments are partly cracked and show bent polysynthetic twins (b, XPL). Original grain size has been extremely reduced by heterogeneous brittle deformation (c, XPL). Larger fragments display an elongated fish-shape, characteristic in shear-deformed cataclastic rocks, and are welded together by cryptocrystalline quartz.

Reactivity potential attributed to:
Strongly strained quartz. Abundant microcrystalline to cryptocrystalline quartz.

Aggregate particle in a sample of concrete prism (Australia).

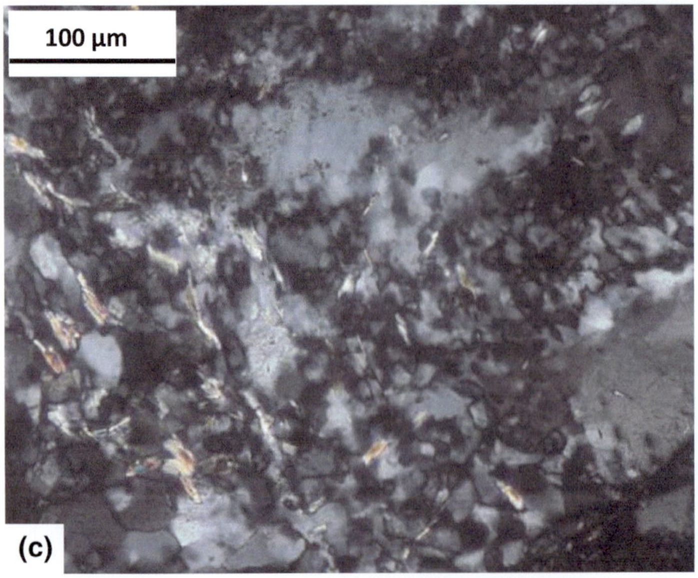

Cataclasite (CH)

Cataclasite composed of quartz, plagioclase, biotite, chlorite, sericite and iron oxides. This tectonite is characterized by brittle and intercrystalline deformation features. There is a large range in grain size and angular outlines of the quartz and feldspar fragments (a, XPL). The larger grains show patchy undulatory extinction. Intense microfracturing and subgraining of quartz which occupy the intergranular spaces are observed. Feldspar shows brittle deformation with multiple fracturing and frictional sliding. Twinning planes of plagioclase are bent and crystals broken into fragments separated by fractures with subgrains of quartz (b, XPL). Microcrystalline to cryptocrystalline subgrains of quartz are abundant in all sectors of the sample and fill the intergranular spaces. Phyllosilicates occur in a sector of the sample and they define a preferred direction of minerals that is parallel to the elongation of the feldspar crystals (c, XPL).

Reactivity potential attributed to:
Strongly strained quartz with recrystallization and subgrains. Microcrystalline to cryptocrystalline quartz is abundant in intergranular spaces.

Aggregate particle in a sample of concrete from a bridge (Switzerland).

Cataclasite (CN)

Cataclasite composed of quartz, plagioclase, biotite, muscovite, titanite, opaque minerals and calcite. The rock is highly deformed and sheared. There is a large range in grain size. The grains (of quartz and feldspar) exhibit angular outlines with straight to sharp boundaries (a, XPL). The rock is intensely cracked (b, XPL) and there is multiple fracturing in the microcrystalline quartz matrix. The crystals of quartz and plagioclase are elongated and stretched. Recrystallization and subgraining generated abundant microcrystalline quartz along microcracks and intergranular spaces (c, XPL).

Reactivity potential attributed to:
Microcrystalline to cryptocrystalline quartz due to strain and fracturing. Deformed crystals of quartz with sutured boundaries. Recrystallization and subgraining.

Sample of rock classified as potentially reactive in laboratory tests (China).

Cataclasite (NO)

Cataclasite composed of quartz, feldspar, chlorite and epidote. Fine-grained rock in which feldspar and quartz porphyroclasts are surrounded by abundant microcrystalline to cryptocrystalline quartz as well as some epidote and chlorite (a, XPL). The rock underwent brittle deformation and contains randomly distributed and oriented grains of angular shapes inherited from the original rock. Quartz crystals show elongation, intense undulatory extinction, sutured boundaries and subgrains (b, XPL). There are some cracks perpendicular to elongation and relicts of deformed quartz grains and subgrains with rotation recrystallization. Locally, epidote is abundant and also forms grains moulded against the coarse clasts (c, XPL).

Reactivity potential attributed to:
Stretched coarse grains of quartz with deeply sutured boundaries. Abundant microcrystalline and cryptocrystalline quartz.

Aggregate particle in a concrete prism test specimen. This aggregate performed as reactive in an airport pavement (Norway).

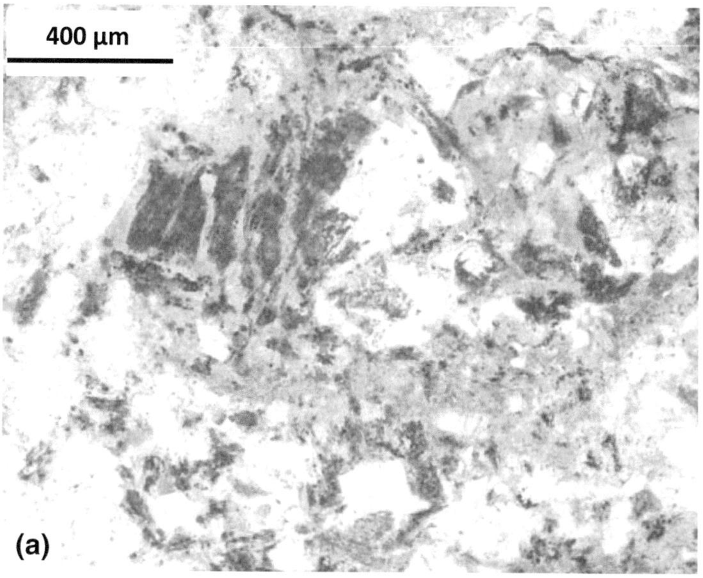

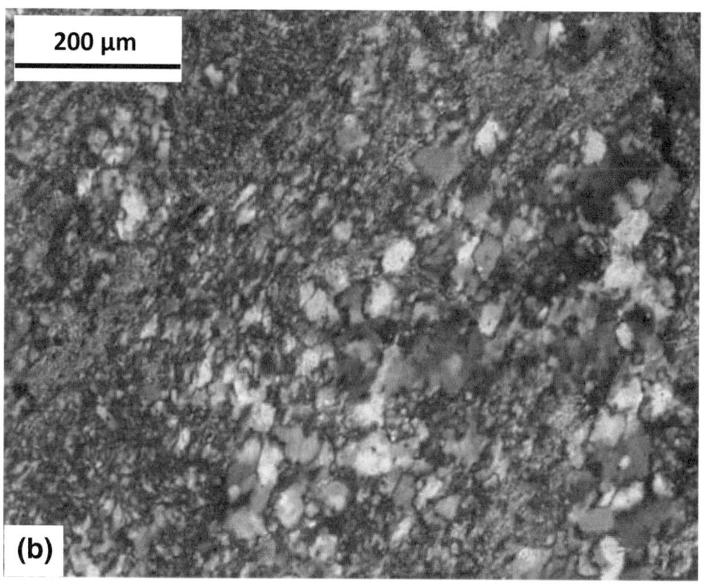

Cataclasite (SE)

Cataclasite composed of quartz, biotite, chlorite and sericite. This tectonite was formed under brittle regime and underwent a later hydrothermal alteration resulting in the formation of the secondary minerals, sericite and chlorite (brownish) (a, PPL). Stretched quartz crystals with evidence of dynamic recrystallization and irregular shapes of grains define preferred orientation. The sericite clusters developed parallel to this orientation (b, XPL). There are polycrystalline rock fragments comprising quartz with subgrains, undulatory extinction and strongly sutured boundaries. In some sectors, quartz occurs as microcrystalline to cryptocrystalline agglomerations (c, XPL).

Reactivity potential attributed to:
Crystals of stretched quartz with sutured boundaries. Abundant microcrystalline to cryptocrystalline quartz.

Aggregate particle in a concrete prism test specimen (Sweden).

Fault breccia (DE)

Fault breccia composed of quartz, feldspar, sericite, chlorite, rutile, titanite, iron oxides and opaque minerals. This sample corresponds to a fault zone of a rock composed mainly of quartz and feldspar. There is subgraining and recrystallization and, in some sectors, a preferred orientation of the crystals (a, XPL). The groundmass is brecciated showing cracks and fragments of minerals with sharp outlines (b, XPL). The feldspar porphyroclasts are slightly altered to clay minerals and sericite. The groundmass is cryptocrystalline and may contain amorphous silica (isotropic) (b, c, XPL).

Reactivity potential attributed to:
Microcrystalline to cryptocrystalline quartz.
Amorphous silica in the groundmass.

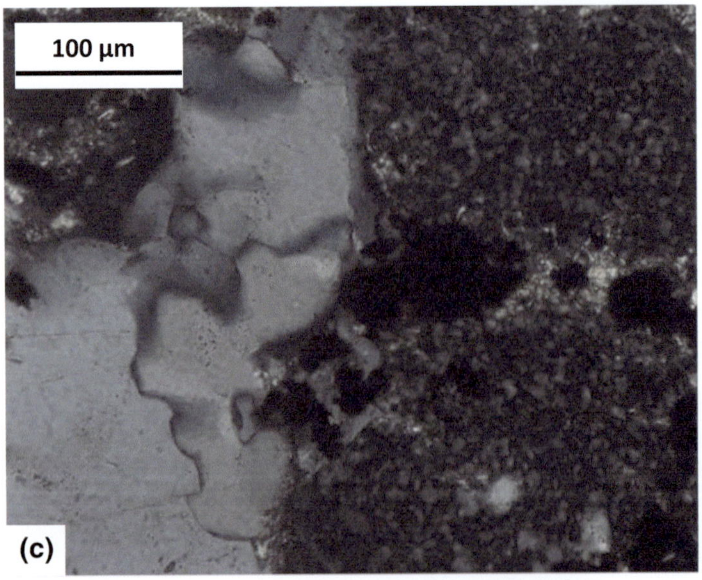

Aggregate particle in a sample of concrete from a hydraulic structure (Germany).

Fault breccia (SE)

Fault breccia rock composed of quartz, K-feldspar, plagioclase, biotite, sericite, opaque minerals and calcite. This recrystallized tectonite is a quartz-feldspathic rock with formation of secondary sericite and calcite, which gives a cloudy tonality to the feldspars (a, XPL). It is characterized by a large range in grain size and strong brittle deformation. The grains exhibit sharp outlines. Larger crystals show straight boundaries. Quartz occurs in developed crystals, in agglomerations of subgrains and also in intergrowths of quartz and K-feldspar by recrystallization (b, XPL). In this case, microcrystalline to cryptocrystalline quartz and feldspar crystals are observed. There are cracks, some filled by secondary quartz growing in comb-shape crystals (c, XPL).

Reactivity potential attributed to:
Microcrystalline and cryptocrystalline quartz, subgrains and intergrowths.

Aggregate particle in a concrete prism test specimen (Sweden).

Fault breccia (UK)

Tectonite composed of quartz, plagioclase limonite and opaque minerals. In this fault breccia brittle deformation is reflected by shear zones observed in the coarse grains (a, XPL). The rock presents a wide range of grain sizes. The grains have angular shapes with straight to sharp edges. The matrix is abundant and it is composed of microcrystalline to cryptocrystalline quartz (b, XPL). The elongated and stretched crystals of quartz and plagioclase exhibit strain lamellae. Fluid inclusion trails are observed in the larger quartz grains. Recrystallization and subgraining generated abundant microcrystalline quartz along microcracks and within intergranular spaces (c, XPL).

Reactivity potential attributed to:
Microcrystalline to cryptocrystalline quartz in the matrix. Deformed crystals of quartz with sutured boundaries, strain lamellae and subgraining.

Aggregate particle in a sample of concrete from a bridge (United Kingdom).

Gneiss (BR)

Gneiss composed of quartz, muscovite, biotite, chlorite, K-feldspar (microcline), plagioclase, epidote, zircon, apatite, opaque minerals and iron oxides. In this rock, deformation partitioning resulted in the division into less strained domains with porphyroblasts/porphyroclasts of microcline and quartz segregated from an intensely strained phyllosilicate-rich domain of smaller grains size (a, XPL). Foliation is formed by the alignment of thin platy phyllosilicates (muscovite and chloritised biotite). Agglomerations of quartz with sutured boundaries are abundant. Very often, deformed-stretched quartz is moulded against the feldspar grains, with C-S fabric (b, XPL) with the most common distribution of elongated crystals of quartz perpendicular or oblique to the C-foliation (b and c, XPL).

Reactivity potential attributed to:
Crystals of strained quartz with alignments of microcrystalline and recrystallized quartz.

Sample of rock used as aggregate in concrete building foundations (Brazil).

Gneiss (CH)

Gneiss composed of quartz, plagioclase, biotite, epidote, chlorite and sericite. The gneissose structure is observed in the quartz- and feldspar-rich layers segregated out from more micaceous layers. The phyllosilicates (biotite and chlorite) plates define a preferred direction of the rock and are bent around the blasts of the harder minerals such as the feldspars (a, XPL). There are also agglomerations of platy crystals in the intergranular spaces. Quartz occurs dominantly in agglomerations of crystals with polygonal boundaries resulting from the recrystallization under metamorphic conditions (b, XPL). Plagioclase blasts show alteration (c, XPL).

Reactivity potential attributed to:
Mosaics of polygonal-shaped recrystallized quartz. Quartz is dominantly microcrystalline.

Aggregate particle in a sample of concrete from a bridge (Switzerland).

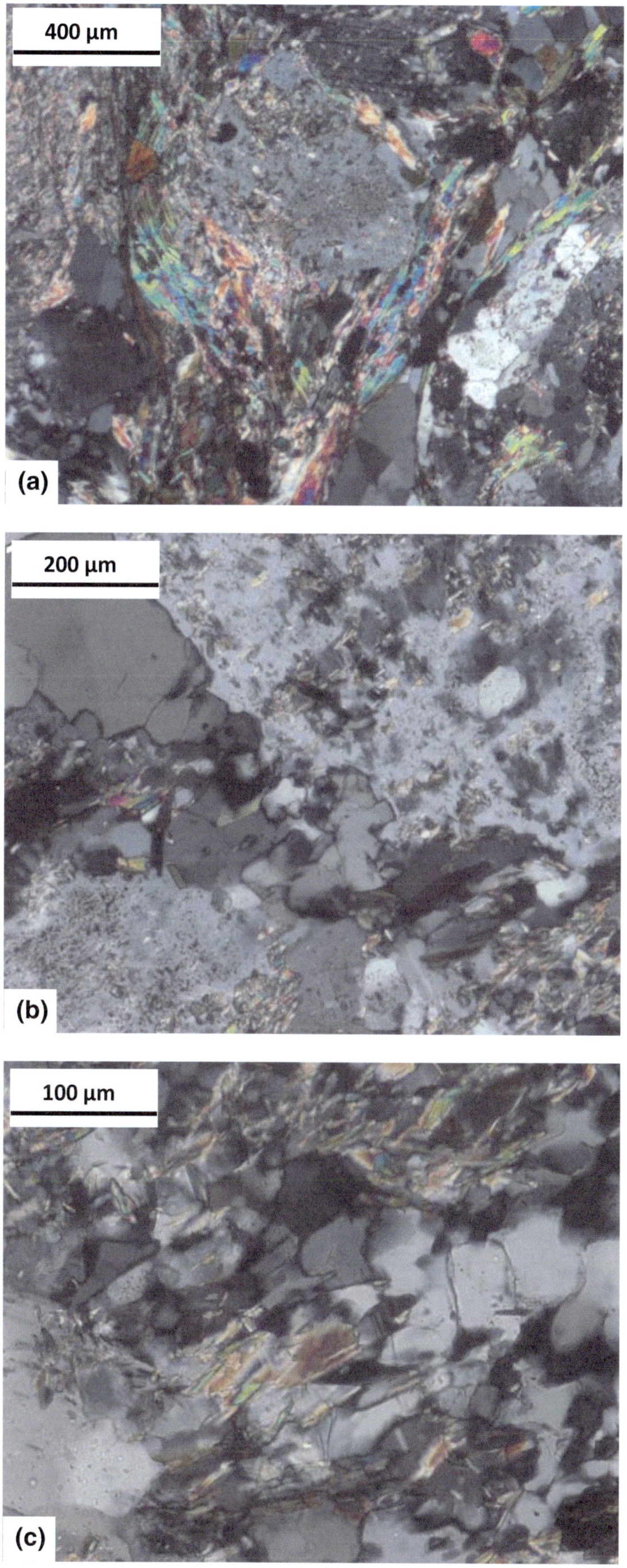

Gneiss (CH)

Heterogeneous granitic gneiss composed of quartz, K-feldspar, plagioclase, muscovite, biotite, chlorite, sericite, garnet, epidote and iron oxides. The rock exhibits preferred orientation of the minerals which are arranged in layers. The mica plates are moulded around the porphyroblasts/porphyroclasts of feldspar which are the larger crystals in the rock (a, XPL). Some of them show rotation and are slightly altered to clay minerals and sericite. Quartz occurs in different sizes and shapes. The larger crystals show undulatory extinction. The smaller crystals occur in the boundaries of the feldspar crystals and most are microcrystalline with sutured boundaries and bulging (b, XPL). In some domains, quartz exhibits triple junction and polygonal mosaics (c, XPL)

Reactivity potential attributed to:
Strongly strained quartz with recrystallization. Microcrystalline quartz occurs in the intergranular spaces.

Aggregate particle in a sample of concrete from a bridge (Switzerland).

Gneiss (CH)

Gneiss composed of quartz, plagioclase, biotite, chlorite, sericite and iron oxides. The rock exhibits gneissose structure with preferred orientation of the minerals. The layers of phyllosilicates are segregated from the quartz and feldspar-rich domains, which are dominant (a, XPL). Plagioclase shows alteration (cloudy). Quartz is deformed: there are stretched and elongated crystals aligned parallel to the preferred orientation of the minerals. Larger crystals of quartz exhibit strong undulatory extinction and deformation lamellae (b, XPL). These are surrounded by smaller crystals in agglomerations locally with straight edges and triple junction points (c, XPL).

Reactivity potential attributed to:
Strongly strained quartz: some clasts are stretched and show strong deformation lamellae and indented boundaries. There is microcrystalline quartz in the boundaries of the larger crystals.

Aggregate particle in a sample of concrete from a bridge (Switzerland).

Gneiss (CH)

Gneiss composed of quartz, plagioclase, biotite, opaque minerals, calcite, apatite, garnet and rutile. The rock shows gneissose structure and features of intense deformation. Biotite is partially chloritised, has opaque minerals and rutile associated, and defines a preferred orientation. Mica-rich domains show a lepidoblastic texture and are banded with granoblastic layers in which recrystallized quartz is dominant (a, XPL). There is evidence of deformation in quartz, plagioclase and biotite crystals. The larger clasts of the main minerals show undulatory and patchy extinction. Quartz crystals have interpenetrating boundaries and show strain lamellae transverse to the elongation of the quartz crystals (b, XPL). There are also bulging, intercrystalline subgrains and sutured boundaries of the quartz crystals (c, XPL).

Reactivity potential attributed to:
Crystals of strained quartz, with strain lamellae and sutured boundaries. Alignments of microcrystalline quartz and subgrains.

Aggregate particle in a sample of concrete from a bridge (Switzerland).

Gneiss (CH)

Fine-grained gneiss composed of quartz, muscovite, biotite, garnet, titanite and opaque minerals. The gneissose banding is evident with quartz domains segregated from the phyllosilicate domains. The darker bands correspond to platy crystals of biotite which define the preferred direction of the rock (a, XPL). Quartz occurs mainly in agglomerations of small crystals. These are stretched, deformed and show undulatory extinction and strain lamellae. The boundaries are deeply sutured, showing bulging and subgrains (b, XPL). The phyllosilicate-rich layers are anastomosing around the thicker quartz layers. Quartz grains have been flattened and dynamically recrystallised. The quartz-quartz contacts show sutured boundaries (c, XPL).

Reactivity potential attributed to:
Crystals of strained quartz. There are alignments of microcrystalline quartz and subgrains.

Aggregate particle in a sample of concrete from a dam (Switzerland).

Gneiss (CH)

Calc-silicate gneiss composed of quartz, muscovite, plagioclase, garnet, epidote and opaque minerals. The rock exhibits a gneissose banding with quartz-rich and mica-rich domains, with epidote associated with muscovite. In the epidote-rich domains, prismatic crystals define a preferred orientation (nematoblastic texture) and this foliation moulds the plagioclase blasts and the lenticular agglomerations of quartz (a, XPL). Plagioclase blasts are slightly altered and usually the twinning planes are perpendicular to foliation. Quartz occurs both in mosaics of polygonal shaped crystals and in agglomerations of stretched crystals with strongly indented edges and recrystallization. Microcrystalline quartz is found locally (b and c, XPL).

Reactivity potential attributed to:
Crystals of elongated stretched quartz with indented boundaries. Alignments of microcrystalline and recrystallized quartz.

Aggregate particle in a sample of concrete from a dam (Switzerland).

Gneiss (FI)

Fine-grained biotite gneiss composed of quartz, plagioclase, K-feldspar, biotite, tourmaline, epidote and zircon.
The rock shows gneissose structure with biotite-rich domains segregated from the felsic minerals-rich domains of quartz, plagioclase and K-feldspar. Preferred orientation is defined by the biotite plates (a, PPL). K-feldspar and plagioclase form the porphyroblasts and are slightly altered (cloudy). Quartz occurs in agglomerations of small polygonal crystals with straight to indented boundaries (b, XPL). Small crystals of quartz (10 μm) are common as inclusions both in the K-feldspar blasts and within the larger crystals of quartz (c, XPL).

Reactivity potential attributed to:
Microcrystalline quartz and abundant tiny quartz inclusions.

Aggregate particle in a sample of concrete from a bridge (Finland).

Gneiss (IT)

Quartz-feldspathic metamorphic rock with gneissose structure composed of quartz, K-feldspar, plagioclase, mica and sericite (from alteration). The porphyroblasts of K-feldspar show patchy extinction and anhedral shapes. The matrix is mainly composed of quartz (a, XPL). Coarser quartz grains have sutured boundaries and undulatory extinction. In the boundaries of the quartz-feldspar grains there are agglomerations of microcrystalline to cryptocrystalline quartz (subgrains) (b, XPL). Microcrystalline quartz is dominant in some domains. The crystals are stretched and also show sutured boundaries (c, XPL).

Reactivity potential attributed to:
Stretched crystals of quartz and abundant subgrains (microcrystalline to cryptocrystalline quartz) in the boundaries of the larger crystals.

Aggregate particle in a concrete prism test specimen. Damage with this aggregate was observed in concrete water structures (Italy).

(a)

(b)

(c)

Gneiss (IT)

Gneiss composed of quartz, epidote, garnet, cordierite, plagioclase, chlorite and iron hydroxides. The rock presents inequigranular granolepidoblastic texture. Plagioclase is abundant and occurs in elongated porphyroblasts, some of which are developed parallel to the preferred orientation defined by the micas (a, XPL). Plagioclase porphyroblasts contain abundant inclusions of very small quartz grains and epidote, which are aligned parallel to the foliation and bent due to deformation (b, XPL). There are quartz-rich domains forming ribbons and mosaics of small crystals with indented edges and agglomerations of microcrystalline quartz. The mica-rich domains exhibit long plates of muscovite some of which mark the preferred orientation while others are randomly oriented (c, XPL).

Reactivity potential attributed to:
Crystals of strained quartz with indented boundaries. Ribbon and microcrystalline quartz.

Aggregate particle in a concrete prism test specimen. Damage in concrete with this aggregate was observed in water structures (Italy).

Gneiss (IT)

Gneiss composed of quartz, biotite, muscovite and iron hydroxides. Deformed rock in which micas occur in restricted domains with preferred orientation defining the bands segregated from the quartz domains (a, b XPL). The crystals of quartz are elongated parallel to the bands formed by the phyllosilicates. Porphyroclasts or layers of deformed quartz grains show undulatory extinction, deformation lamellae, subgrains, irregular embayments and deeply sutured boundaries with bulges and recrystallized grains (b, XPL). Closer to the mica-rich domains, the quartz shows triple-junctions and polygonal texture (c, XPL).

Reactivity potential attributed to:
Crystals of strained, stretched quartz.

Aggregate particle in a concrete prism test specimen. Damage in concrete with this aggregate was observed in water structures (Italy).

Gneiss (NO)

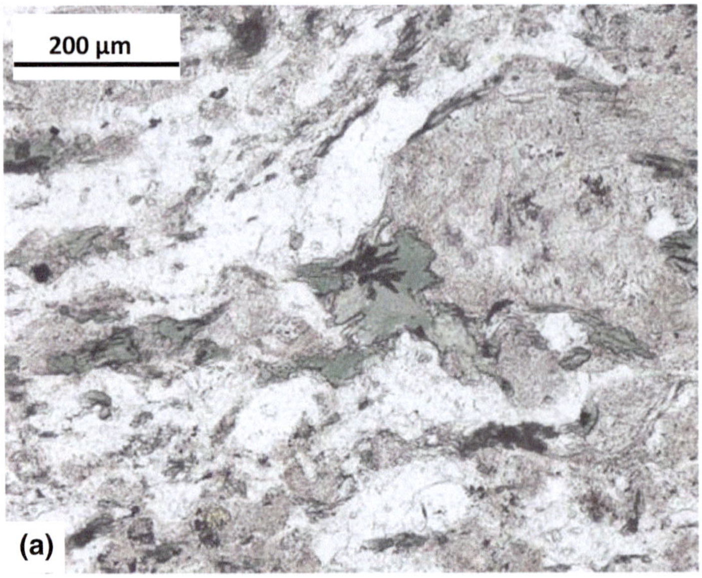

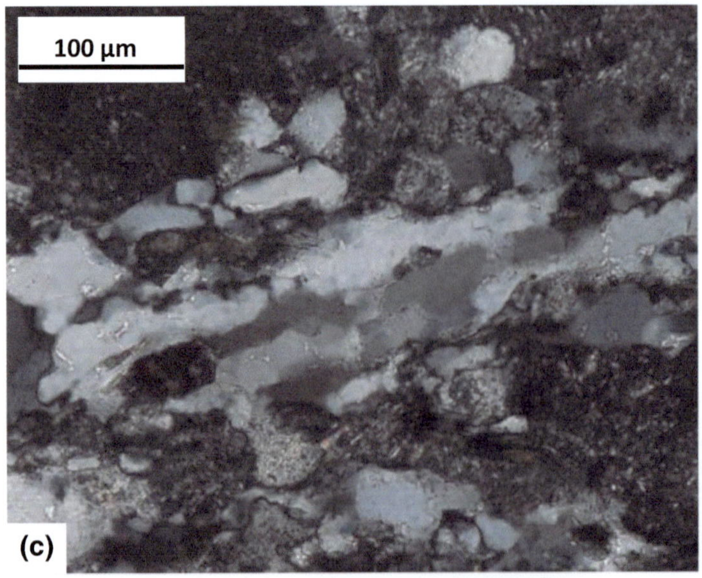

Deformed rock composed of quartz, plagioclase, chlorite, epidote and apatite. This is a deformed, foliated rock that has been affected by hydrothermal alteration. Metamorphic differentiation by deformation is observed in a few strained domains with porphyroblasts of plagioclase surrounded by intensely deformed zones (a, PPL). Feldspar porphyroblasts are strongly altered to sericite. Chlorite forms developed blasts but occurs also in vermicular variety. There are bands of stretched, deformed quartz with strain lamellae, sutured boundaries, subgrains and recrystallization (b, XPL). These bands of agglomerations of quartz crystals develop parallel to the alignment of the compositional banding (c, XPL).

Reactivity potential attributed to:
Crystals of strained quartz with sutured boundaries. Alignments of microcrystalline and subgrained quartz.

Aggregate particle in a concrete prism test specimen. Deleterious ASR with this aggregate was observed in bridges (Norway).

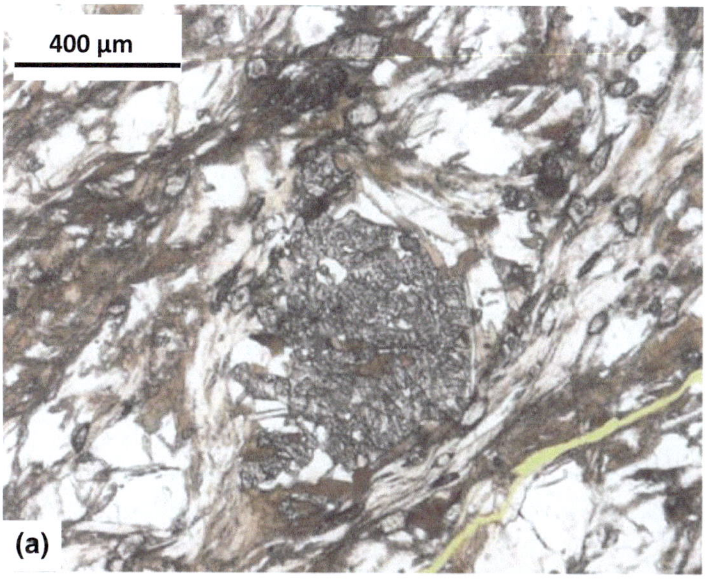

Gneiss (SE)

Gneiss composed of quartz, biotite, garnet, epidote, sericite, plagioclase and titanite. The rock exhibits porphyroblastic-banded texture. The porphyroblasts are formed by poikilitic garnet with abundant inclusions of quartz (a, PPL). The foliation is defined by the preferred orientation of the mica plates. The crystals of quartz are elongated parallel to the foliation (b, XPL), have sutured boundaries and show strong undulatory extinction. There are biotite-muscovite intergrowths associated with the elongated quartz crystals. Subgrains of quartz are common (c, XPL).

Note: the section was mounted using yellow dyed epoxy for fluorescence.

Reactivity potential attributed to:
Abundant microcrystalline quartz. Strained quartz with sutured boundaries and subgrains.

Aggregate particle in sample of concrete from a dam (Sweden).

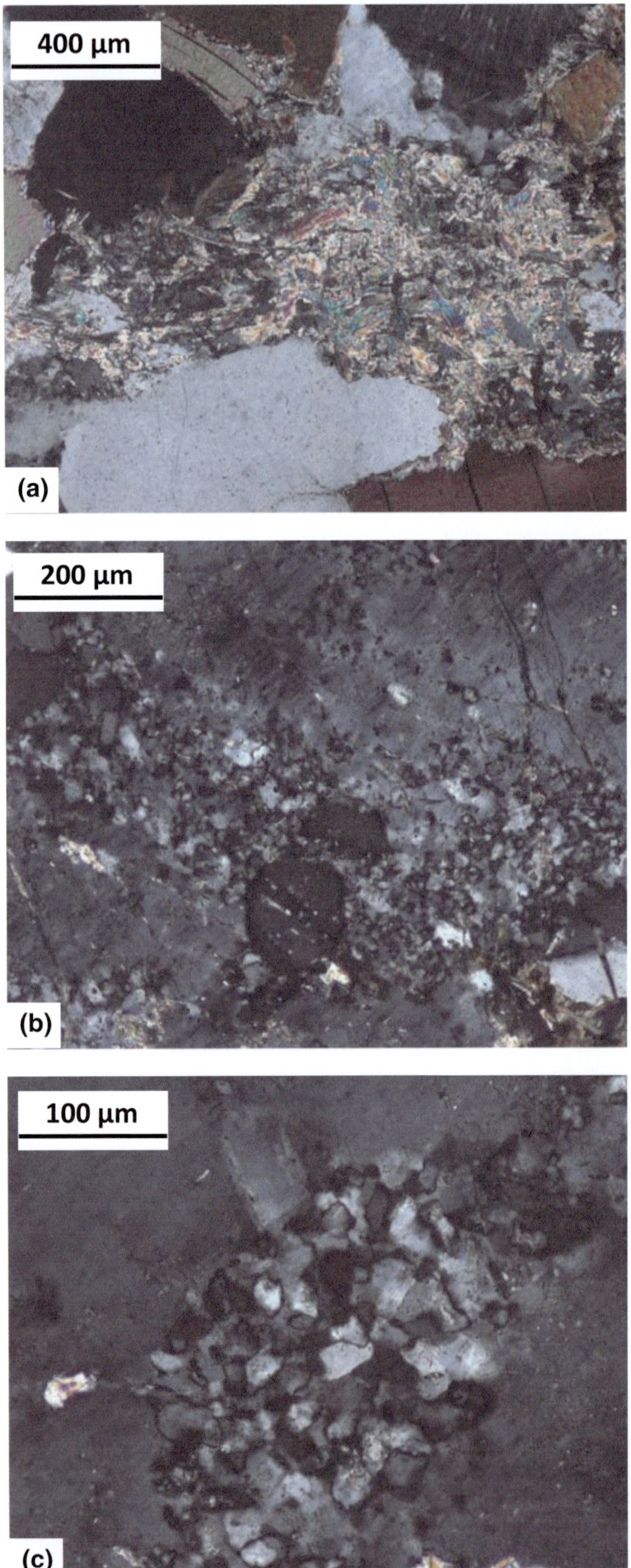

Gneiss (ZA)

Deformed and recrystallized rock of granitic composition containing quartz, perthitic K-feldspar, plagioclase and biotite, with trace amounts of chlorite, sericite, apatite, tourmaline, opaque minerals, and zircon. The rock-forming quartz and K-feldspar are arranged in a heterogranular texture and subordinate mineral species are agglomerated in between. Coarse plagioclase is heavily altered to sericite and chlorite (a, XPL) and contains microcrystalline quartz. Coarse quartz has straight to curved grain boundaries, and commonly displays undulatory extinction and strain lamellae. Deformation concentrates in trails, causing subgraining in coarse grained quartz and K-feldspar (b, XPL). Cracks are frequent. Clusters of subgrained microcrystalline quartz occupy the interstitial space between large K-feldspar crystals (c, XPL).

Reactivity potential attributed to:
Microcrystalline quartz in deformation trails.

Aggregate particle in concrete from a railway sleeper (South Africa).

Granofels (FI)

Granofels composed of quartz, biotite, plagioclase and K-feldspar. This is a very fine-grained isogranular rock with granoblastic texture (a, XPL). The tiny blasts of biotite (brown) are homogeneously distributed, with no preferred orientation (b, PPL). Quartz is dominant and occurs as very small crystals of irregular shapes and indented boundaries (c, XPL). There are rare clasts (quartz grains of the protolithic rock) of slightly larger dimension.

Reactivity potential attributed to:
Microcrystalline quartz is abundant and shows sutured borders. Crystals are interlocked in each other.

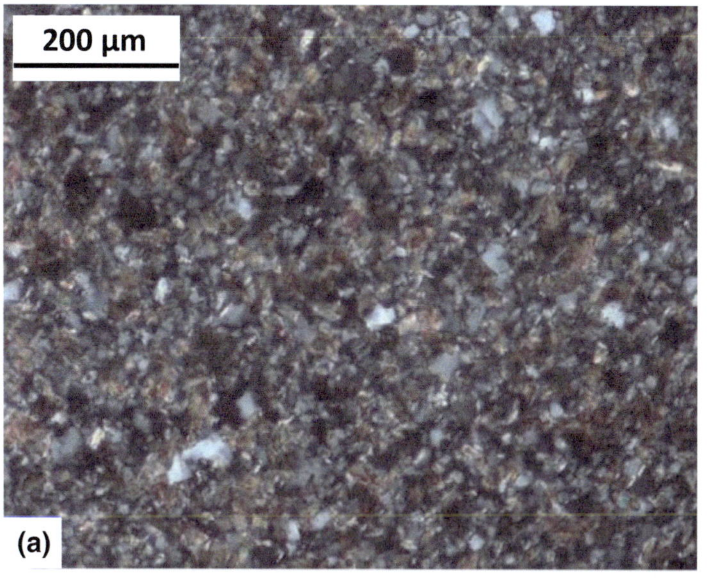

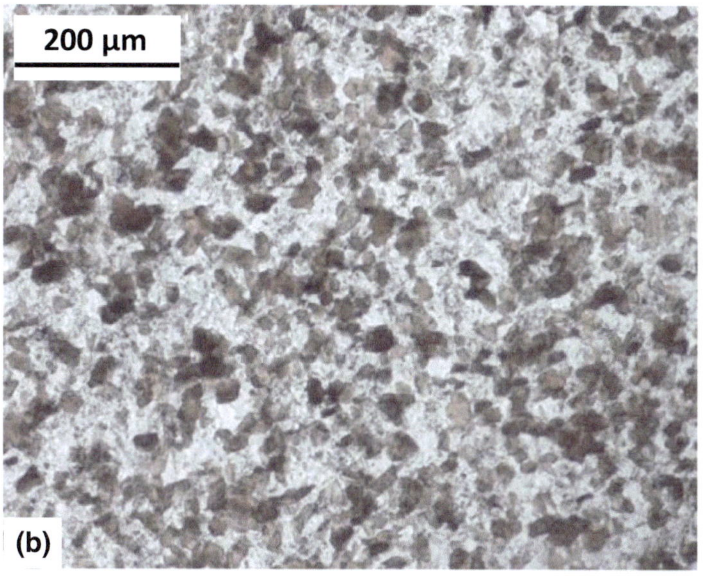

Aggregate particle in a sample of concrete from a bridge (Finland).

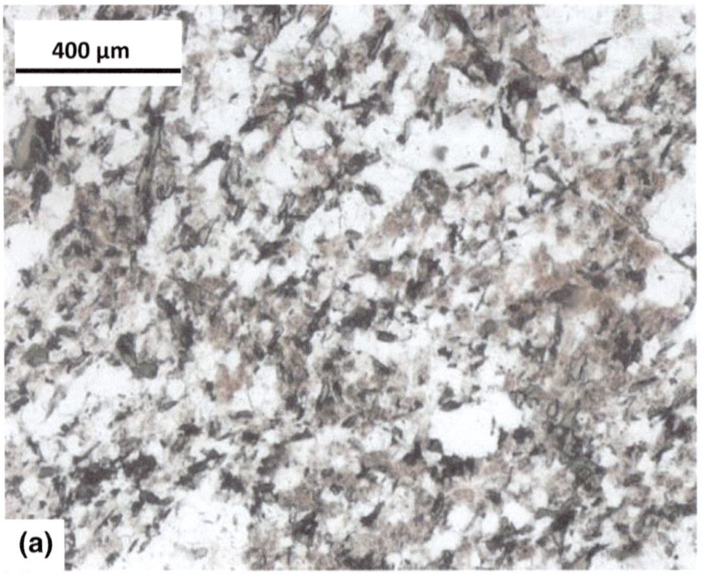

Granofels (FI)

Granofels composed of quartz, biotite, plagioclase, chlorite and sericite. The rock is slightly inequigranular with granoblastic texture and weak preferred orientation of the phyllosilicates. The blasts of biotite are tabular, slim and show partial to total chloritization (a, PPL, greenish). Quartz occurs as small crystals of irregular shapes and also as agglomerations of polygonal crystals (b, XPL). The larger crystals of quartz exhibit undulatory extinction and sutured limits. The larger clasts (protolithic minerals) in the rock are of feldspar partially altered (cloudy). These blasts contain abundant inclusions of tiny rounded (cryptocrystalline) quartz grains (c, XPL).

Reactivity potential attributed to:
Microcrystalline quartz is dominant and exhibits indented boundaries. There is also cryptocrystalline quartz in inclusions in the feldspars.

Aggregate particle in a sample of concrete from a bridge (Finland).

Granofels (FI)

Granofels composed of quartz, biotite, chlorite, muscovite and K-feldspar. This is a fine-grained rock with granoblastic texture in which quartz and biotite occur in similar proportion. The blasts of both minerals are randomly distributed, although there is a weak preferred orientation of the lamellar blasts (a, PPL). The matrix contains cryptocrystalline quartz (b, XPL) but the identification of the components is difficult owing to the small dimensions of the crystals. Higher magnification (c, XPL) shows a matrix composed of microcrystalline to cryptocrystalline quartz.

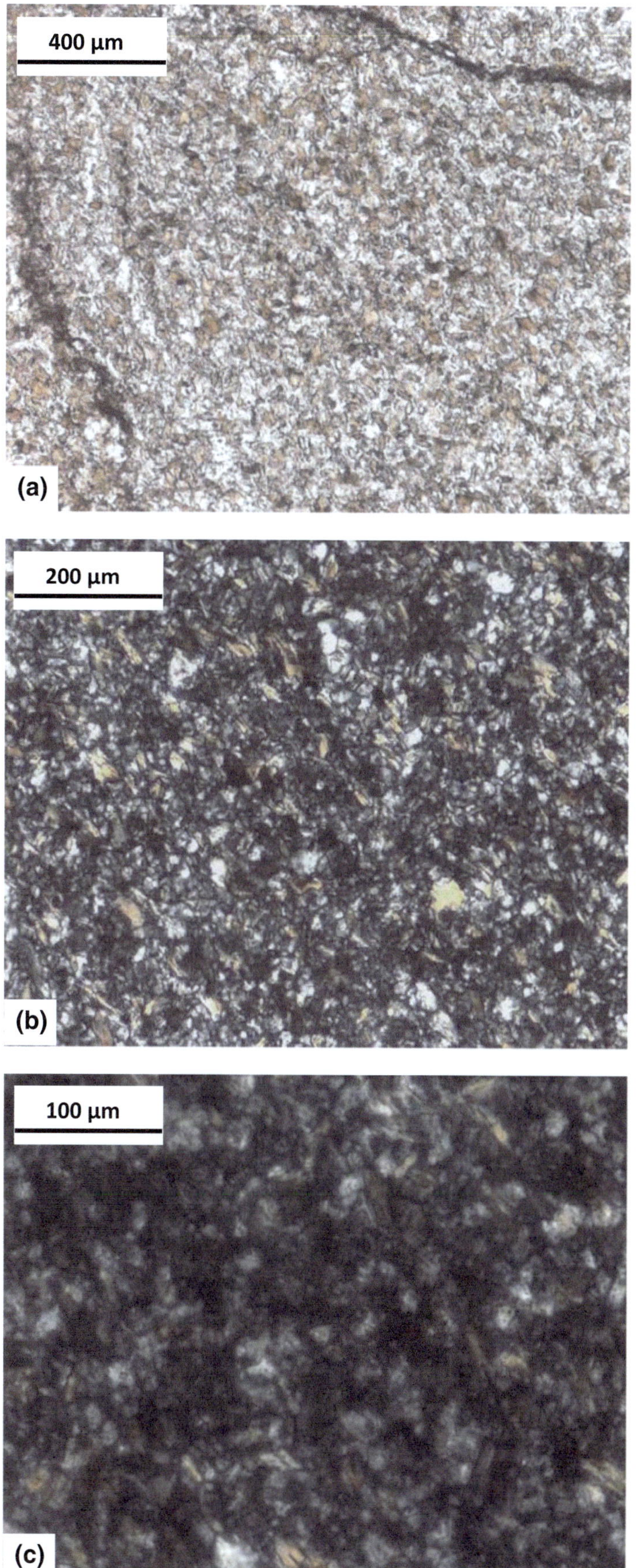

Reactivity potential attributed to:
Microcrystalline to cryptocrystalline quartz is the main component of this granofels. Intergrowth in a random interlocking pattern of abundant quartz and biotite is observed at high magnification.

Aggregate particle in a sample of concrete from a bridge (Finland).

Granofels (NA)

Granofels composed of quartz, K-feldspar, orthopyroxene, clinopyroxene, amphibole, garnet, epidote, titanite, calcite, zircon, apatite and opaque minerals. The rock is characterised by a granoblastic texture in which domains with triple-junctions can be observed. The rock shows abundant microfractures, mainly in the orthopyroxene blasts which exhibit alteration along cracks (a, XPL). Quartz is abundant and forms some blasts larger than the matrix of the rock. However, there are also some tiny crystals of quartz included in the larger crystals of feldspar. Static recrystallization with grain boundary area reduction is evident by the triple-junctions of the quartz crystals. Quartz forms anhedral grains with straight to convex boundaries (b, XPL) and undulatory extinction angle. Myrmekites are common (c, XPL).

Reactivity potential attributed to:
Quartz micro-inclusions, myrmekite, microcrystalline quartz as alteration by-product.

Aggregate particle in concrete from an electric mast foundation (Republic of Namibia).

Granofels (USA)

Inequigranular coarser-grained granofels composed of quartz, kyanite and opaque minerals. The rock is composed mainly of quartz formed under high pressure metamorphism of a quartz-rich protolith. The texture is mainly polygonal with triple-junctions (a, XPL). The crystals are elongated defining a slight preferred orientation with kyanite blasts parallel to this direction (b, XPL). The boundaries of the quartz crystals are mainly straight (c, XPL).

Reactivity potential attributed to: *Features of pressure-solution of the protolith rock. Elongated crystals of quartz. Abundant recrystallized quartz with polygonal boundaries.*

Aggregate particle in a concrete prism test specimen (USA).

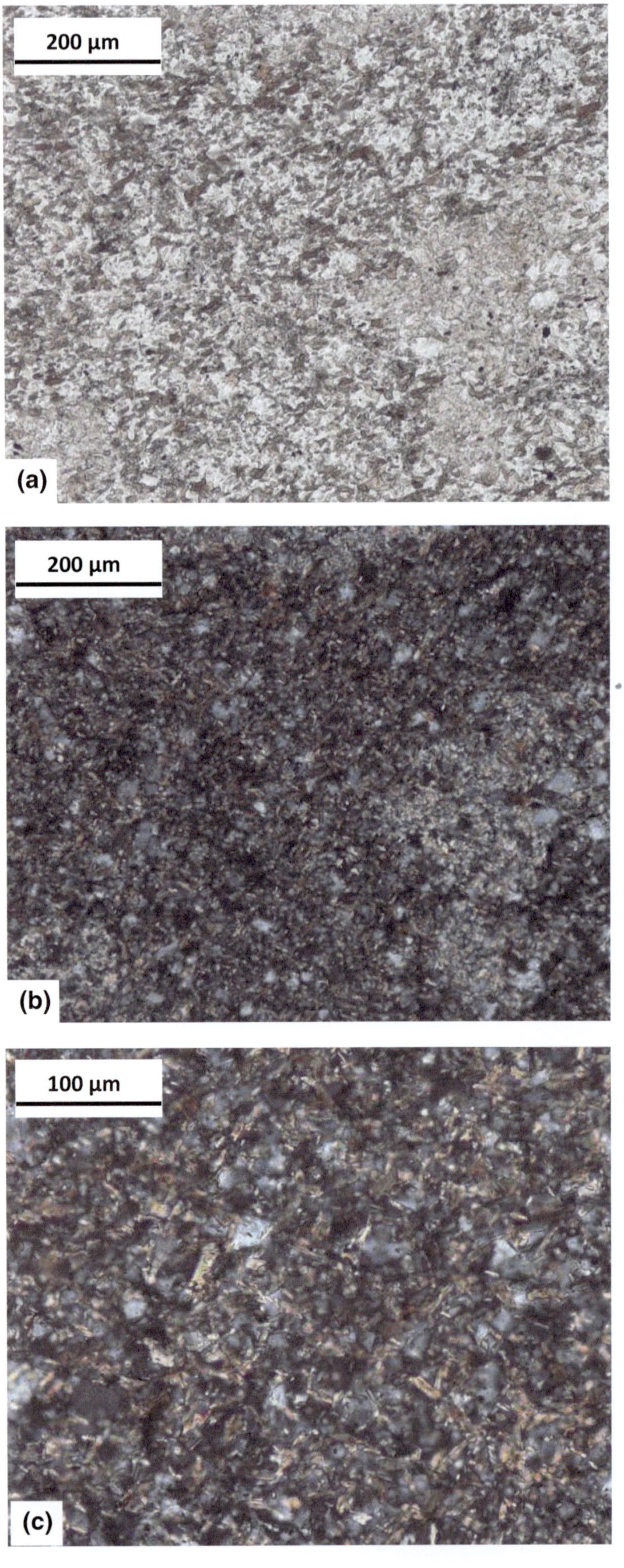

Granofels (ZA)

Fine-grained granofels composed of quartz, biotite, cordierite, chlorite, with minor sericite, and traces of tourmaline, epidote, titanite, and opaque minerals, arranged in a random fabric. There is no preferred orientation (a, PPL; b, XPL) of the minerals although the sedimentary layering is still visible. Quartz and biotite show similar content. The lighter spots correspond to anhedral poikilitic blasts of cordierite (a, PPL; b, XPL). At high magnifications, biotite flakes are seen to form a reticulate network (c, XPL). Due to its very-fine grained nature and mineral content, grain and boundary shape of quartz are difficult to assess.

Reactivity potential attributed to:
Microcrystalline to cryptocrystalline quartz ubiquitous in the rock.

Aggregate sample from a formation classified as potentially reactive (South Africa).

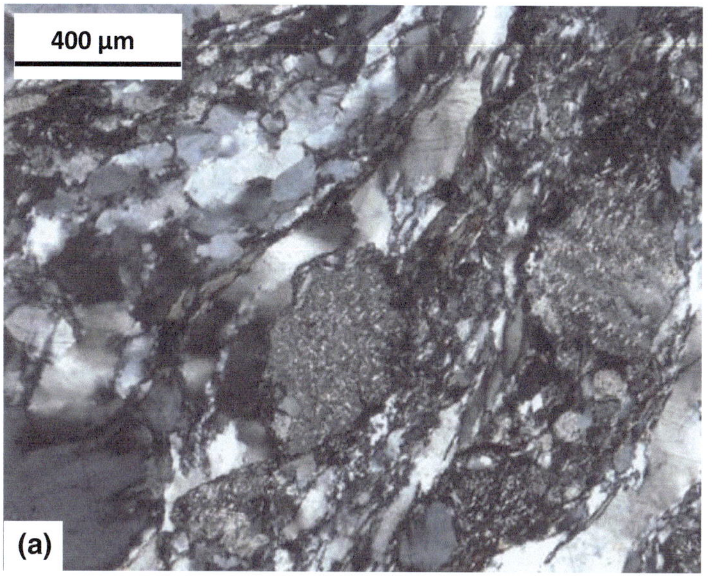

Mylonite (FI)

Intensely deformed metasedimentary rock composed of quartz, K-feldspar, chlorite and sericite. The rock presents mylonitic texture. There is local preferred orientation of the minerals where the scarce chlorite plates and the more intense deformation bands with subgraining of quartz surround the grains of K-feldspar (a, XPL). Perthitic partially altered K-feldspar forms the larger crystals. Crystals of quartz show bulging, inclusion trails and deformation lamellae (b, XPL). Quartz also occurs as coarse elongated and stretched grains. The boundaries are strongly indented. In the boundaries there are abundant subgrains with microcrystalline to cryptocrystalline quartz and feldspar (c, XPL).

Reactivity potential attributed to:
Stretched quartz with sutured boundaries and deformation lamellae. The boundaries of the larger blasts are lined by abundant subgrains.

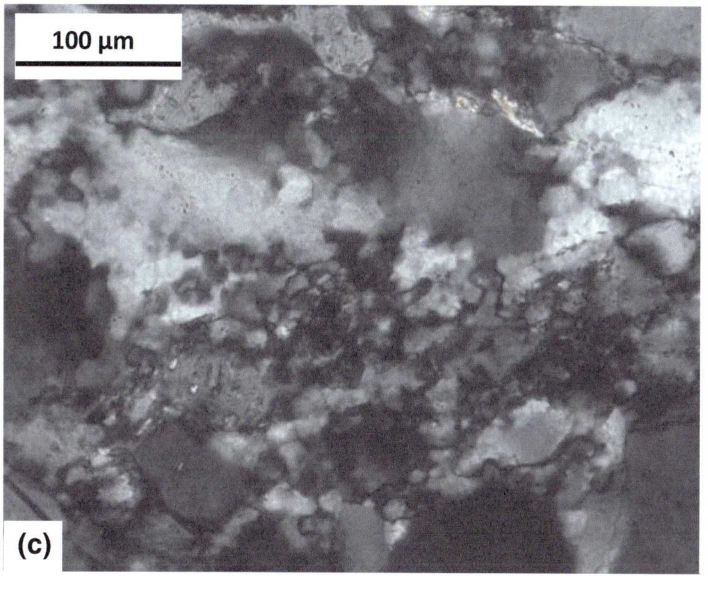

Aggregate particle in a sample of concrete from a bridge (Finland).

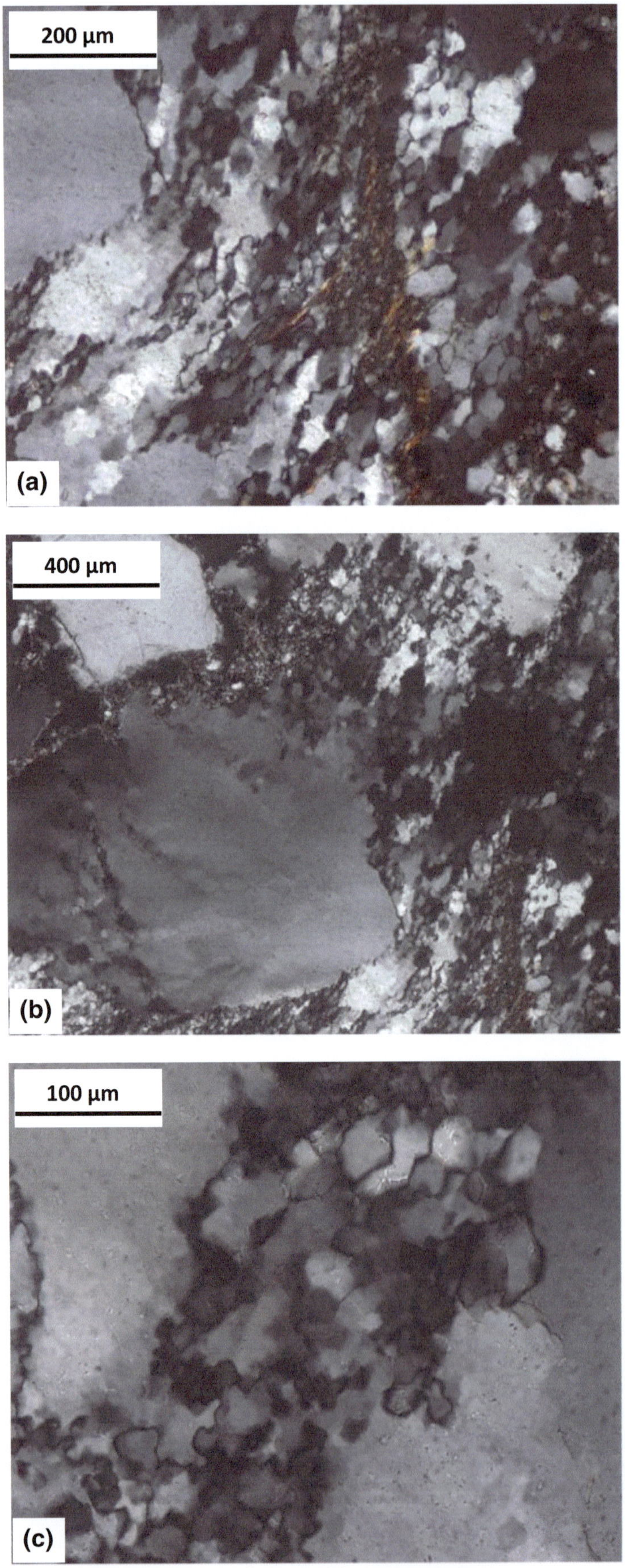

Mylonite (JP)

Mylonite composed of quartz and chlorite. The rock exhibits features of dynamic subgraining and recrystallization. There is deformation partitioning with larger grains less deformed than the surrounding small crystals (a, XPL). The boundaries of the grains are deeply serrated and interlobate. The crystals exhibit deformation lamellae and inclusion trails. There are intracrystalline shear bands in the coarser quartz crystals (b, XPL). Recrystallization and subgraining are dominant along serrated grain boundaries (c, XPL). Abundant ribbon quartz occupies the spaces between coarse grains and defines the preferred orientation of the rock, locally marked by plates of chlorite.

Reactivity potential attributed to:
Crystals of strained quartz with deformation lamellae. The boundaries of the crystals are intensely indented. There is abundant microcrystalline to cryptocrystalline quartz.

Aggregate particle in a concrete prism test specimen (Japan).

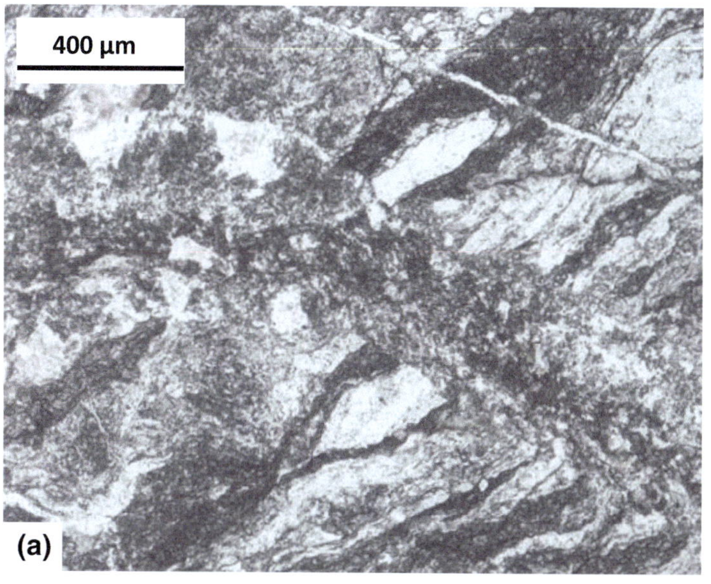

Mylonite (NO)

Mylonite composed of quartz, plagioclase, epidote, sericite, chlorite and titanite.
This rock with calc-silicate composition was formed by mechanical reduction of grain size due to ductile deformation. It is strongly strained with recrystallization of minerals, mainly of quartz (a, PPL). Feldspar porphyroclasts remained as large relict grains, although with strongly fractured margins (b, XPL). There are clots of epidote in agglomerations and also as euhedral crystals. Shear zones at the microscopic scale with fine granoblastic quartz grains and a fine foliation associated with stretched crystals are observed. The deformation did not destroy totally the original fabric. Quartz has undergone dynamic recrystallization and subgraining giving rise to a fine-grained matrix of undeformed grains. These grain boundaries show embayments (c, XPL).

Reactivity potential attributed to:
Alignments of microcrystalline quartz.
Abundant subgrained quartz.

Aggregate particle in a concrete prism test specimen (Norway).

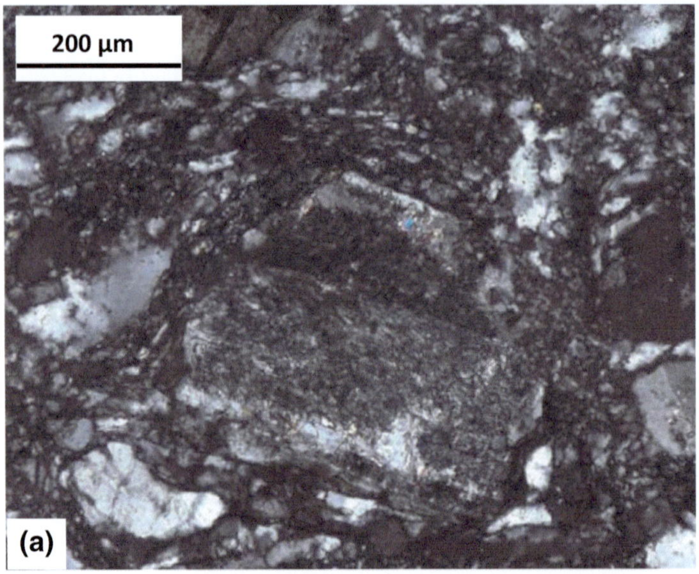

Mylonite (NO)

Mylonite composed of quartz, plagioclase, epidote, sericite, chlorite and titanite. This is a strained rock of calc-silicate composition which resulted in the reduction of grain size (subgrains) by ductile to brittle deformation (a, XPL). There is a mylonitic texture with a penetrative foliation and alignment of the stretched crystals such as those of quartz (b, XPL). Some porphyroblasts are preserved although deformed and are surrounded by microcrystalline quartz. There is strong grain reduction in virtue of ductile deformation. Quartz crystals are stretched, show sutured boundaries and subgrains (c, XPL).

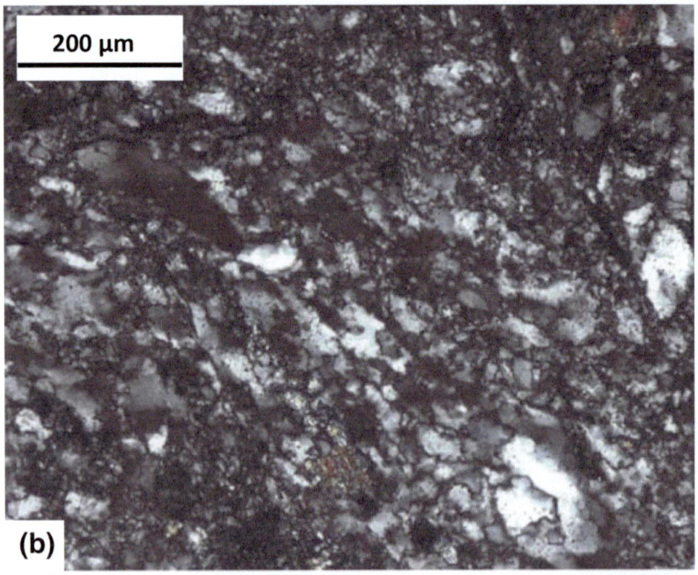

Reactivity potential attributed to:
Alignments of microcrystalline quartz, stretched crystals of quartz with indented boundaries and subgraining.

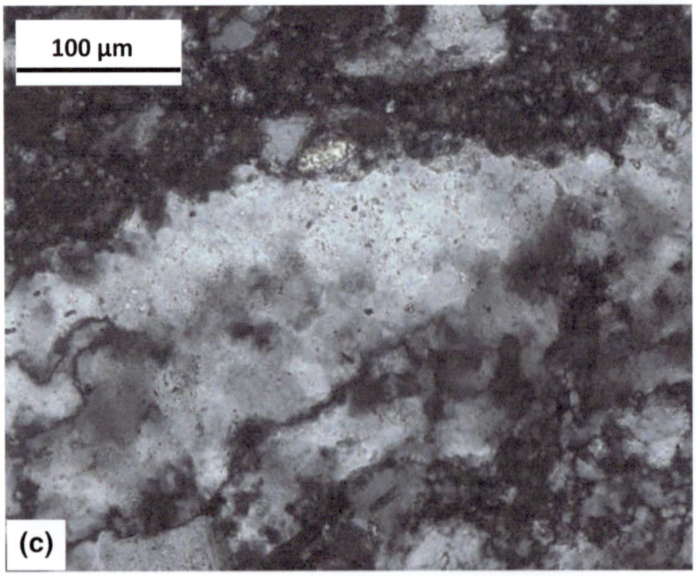

Aggregate particle in a concrete prism test specimen (Norway).

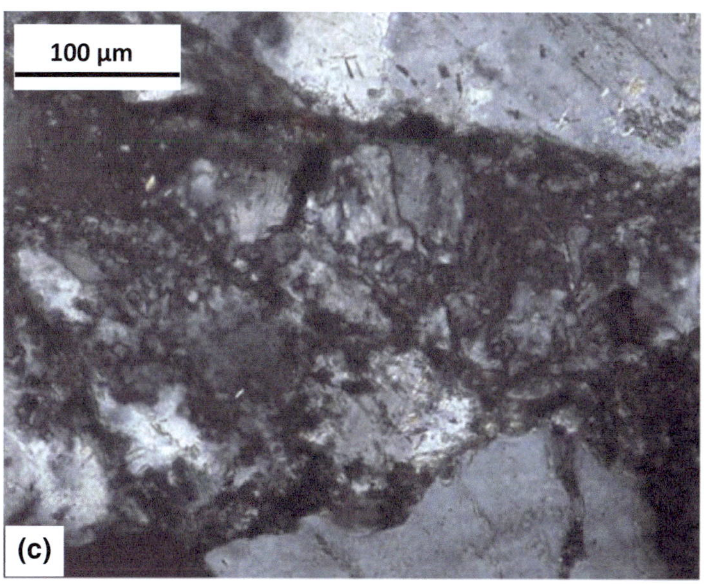

Mylonite (SE)

Mylonite composed of quartz, K-feldspar, plagioclase, chlorite and opaque minerals. The rock exhibits a protomylonitic texture due to ductile deformation (a, XPL). Quartz is the main component of the rock and occurs as developed grains and also in subgrains. Biotite crystals and chlorite develop in the intergranular spaces. The larger grains of quartz are cloudy owing to deformation lamellae and abundant fluid inclusion trails. These grains show sutured boundaries and most of them exhibit undulatory extinction. Feldspar crystals show fracturing and patchy extinction. The intergranular spaces are mainly filled by intergrowths of feldspar with microcrystalline to cryptocrystalline quartz (b, XPL). Crystals are smaller than 50 µm and the intergranular spaces and cracks are lined by iron hydroxides (c, XPL).

Reactivity potential attributed to:
Crystals of strained quartz. The boundaries of the crystals are intensely indented. There are intergrowths of microcrystalline to cryptocrystalline quartz and feldspar

Aggregate particle in a sample of concrete from a dam (Sweden).

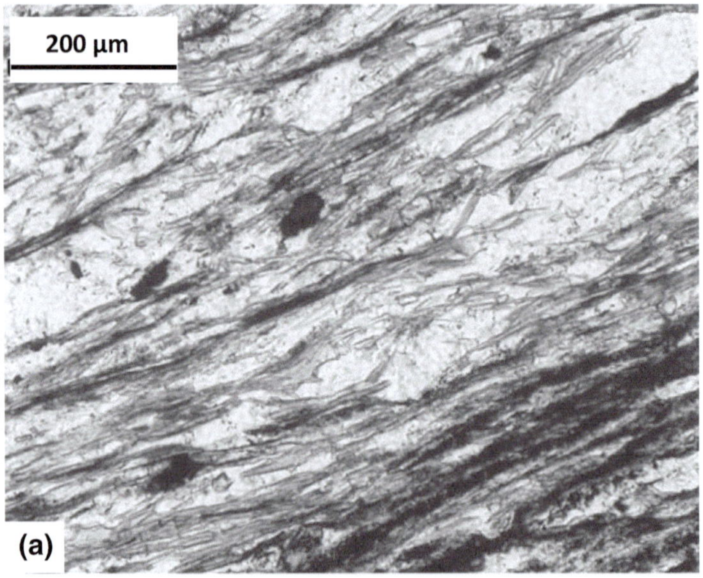

Phyllite (NO)

Quartz-phyllite composed of quartz, muscovite, plagioclase, tourmaline, opaque minerals, chlorite, calcite, organic matter and dolomite (secondary). The rock shows preferred alignment of the phyllosilicates in bladed crystals displaying a perfect penetrative foliation (a, PPL). This is a fine-grained rock of low metamorphic grade, in which the quartz crystals are distributed in lenticular aggregates moulded by the phyllosilicates. Euhedral porphyroblasts of dolomite of late crystallization can be found (b, XPL). Elongated (stretched) quartz crystals, due to pressure-solution, are parallel to the mica preferred orientation. Most of the quartz is microcrystalline and shows straight to indented boundaries (c, XPL).

Reactivity potential attributed to:
Quartz with features of pressure-solution. Stretched quartz, elongated parallel to the foliation. Abundant microcrystalline quartz.

Aggregate particle in a concrete prism test specimen (Norway).

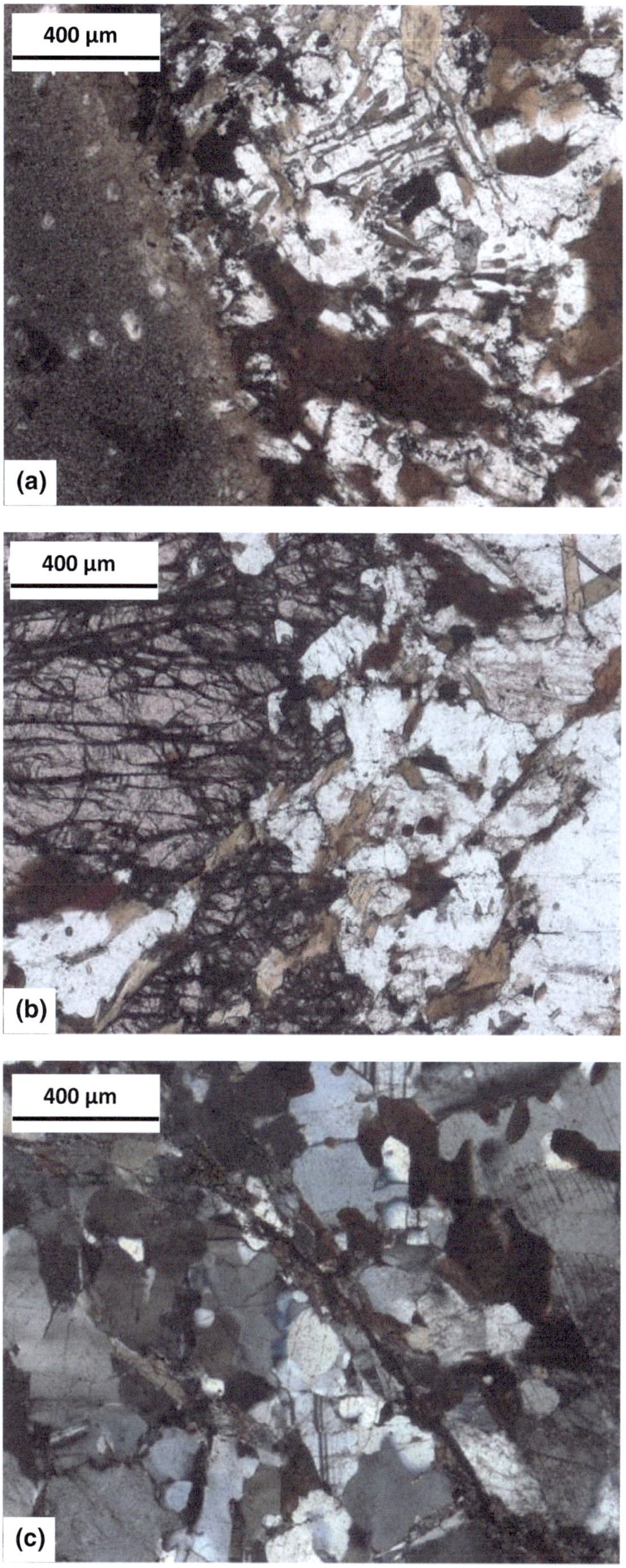

Pseudotachylite (AR)

Pseudotachylite (tectonite with fragmentation associated with partial melting by frictional heating) composed of quartz, K-feldspar, plagioclase, biotite, pyroxene, amphibole, apatite, zircon and garnet. Very fine-grained fault rock with cataclastic texture (brecciated zone) and melted material. It contains cryptocrystalline to vitreous groundmass resulting from frictional sliding with fusion of the materials of the fault gouge (a, PPL). The original rock, probably an orthogneiss, underwent cataclasis and development of pseudotachylitic veins and glassy material. The brittle deformation in the rock generated cracking of a garnet crystal. The original rock shows bands of biotite developing foliation (b, PPL). In this domain the quartz crystals show cracks, undulatory extinction, inclusion trails and sutured boundaries (c, XPL).

Reactivity potential attributed to:
Amorphous to low crystallinity material in brecciated zones. Rock with a high content of glassy material formed by friction melting at high strain rate under a brittle to brittle-ductile regime. Coarser quartz is deeply strained.

Sample of rock used as aggregate in a concrete pavement (Argentina).

Quartzite (AU)

Medium to fine-grained impure quartzite composed of quartz, muscovite and biotite, locally lined with limonite. Coarse quartz grains display undulatory extinction and extensive subgraining along the periphery, further recrystallizing to form a locally equigranular fabric (a, XPL). In the presence of micas (muscovite, biotite), quartz grain shape is a bit more elongate highlighting the incipient foliation (b, XPL). Fluid inclusion trails are seen to cross the present quartz fabric. Quartz-quartz grain contacts range from straight with triple-junctions to slightly concave-convex (c, XPL).

Reactivity potential attributed to:
Intensely strained quartz with interlobate boundaries and subgraining. Abundant microcrystalline quartz.

Aggregate particle in a sample of concrete from a hydraulic structure (Australia).

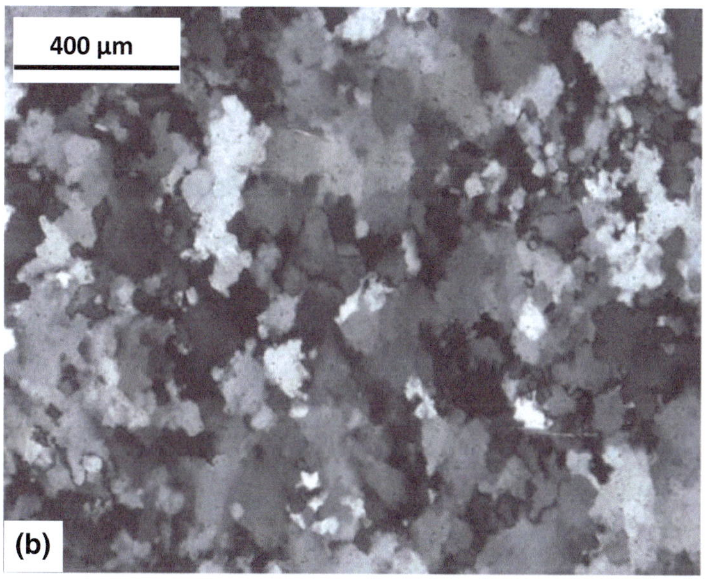

Quartzite (BR)

Quartzite composed just of quartz.
The ductile deformation of the rock is reflected in the deeply serrated edges (a, XPL). The rock is highly deformed and extensively recrystallized with smaller grains along the boundaries of the larger subgrained quartz. There is a weak alignment of the elongated crystals (b, XPL). Deformation and recrystallization of the microstructure of host grains is observed, with bulging and boundary migration (c, XPL).

Reactivity potential attributed to:
Intensely strained quartz with interlobate boundaries and subgraining. Quartz shows bulges, grain boundary migration and recrystallized grains.

Sample of rock used as aggregate in a concrete dam (Brazil).

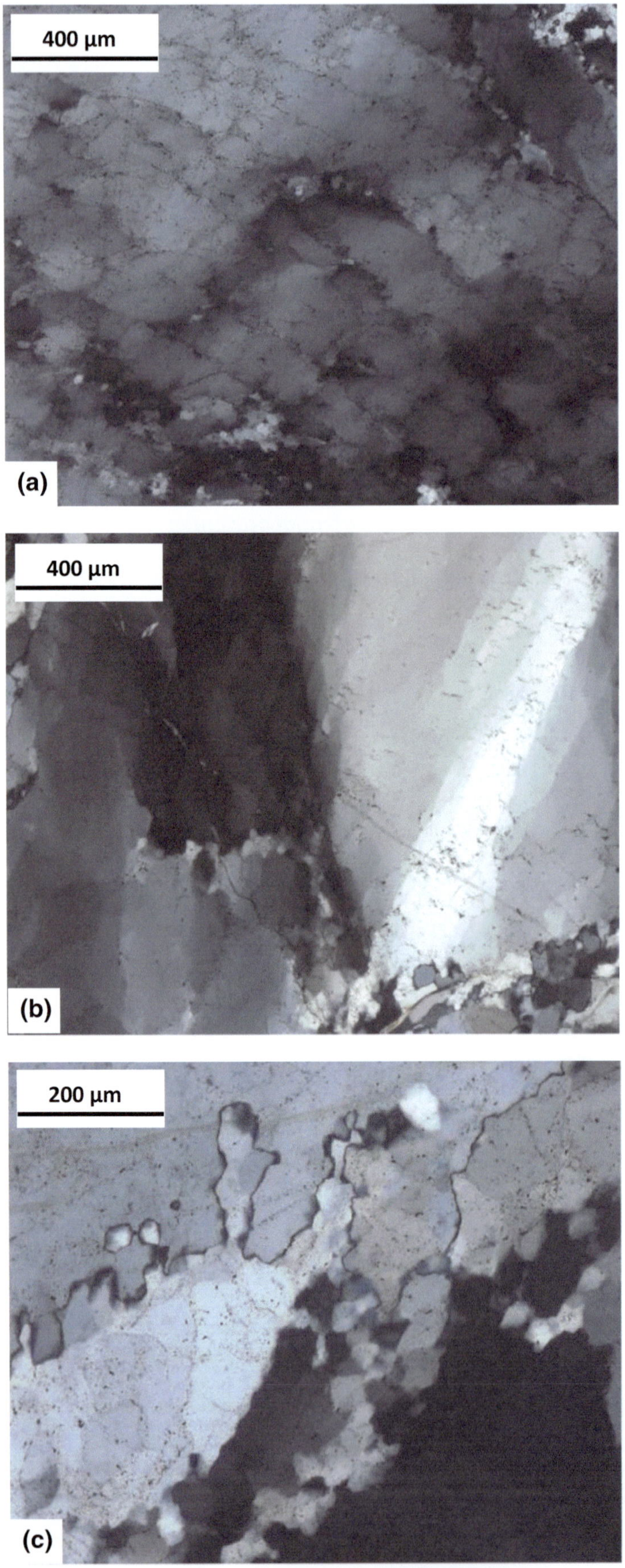

Quartzite (DE)

Deformed rock (quartzite) composed of quartz, chlorite and sericite. This quartzite shows features of deformation with recrystallization by grain boundary migration (a, XPL). It is an intensely strained quartzite with quartz grains showing strong undulatory extinction, parallel deformation bands (i.e., stripes of slightly different extinction position) and subgrains. Large quartz grains with undulatory extinction and elongated subgrains pass laterally into domains of small grains (microcrystalline quartz) (b, XPL). There is bulging recrystallization of quartz. Grain boundaries are irregular due to grain boundary migration and subgrain rotation (c, XPL).

Reactivity potential attributed to:
Intensely strained quartz with interlobate boundaries, bulges, grain boundary migration and microcrystalline quartz.

Aggregate particle in a sample of concrete from a hydraulic structure (Germany).

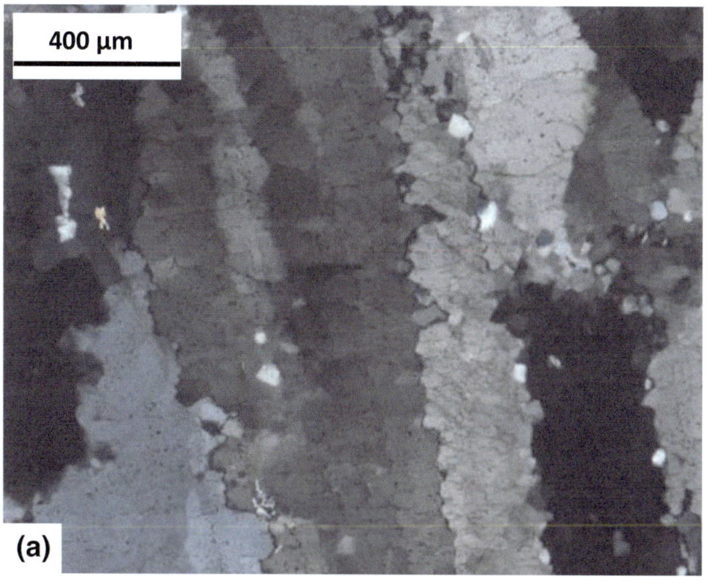

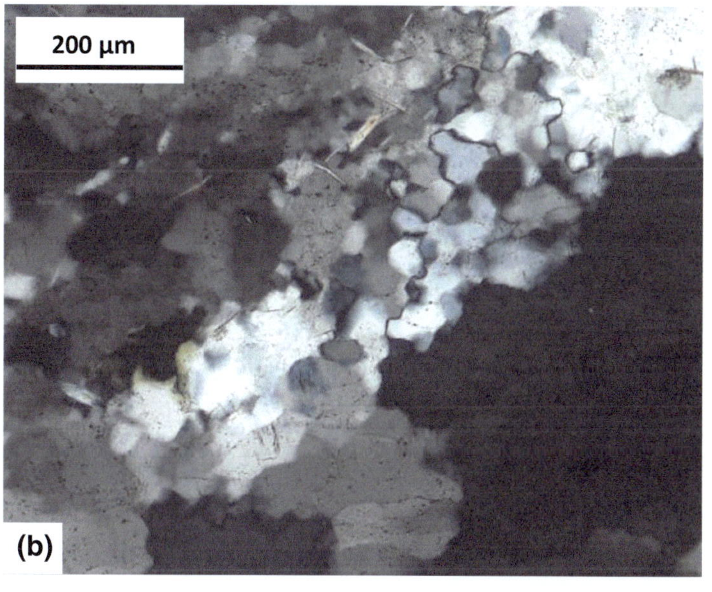

Quartzite (DE)

Intensely deformed rock composed of quartz, chlorite and sericite. This quartzite is intensely strained. It exhibits strong undulatory extinction, stripes of different extinction position (deformation lamellae and ribbon quartz) (a, XPL) and subgrains. These lamellae seem to present chessboard pattern subgrains. In the contacts between large quartz grains there are domains of microcrystalline quartz (b, XPL). There is bulging recrystallization of quartz and the grain boundaries are irregular due to grain boundary migration and subgrain rotation (c, XPL). The bulges show irregular shapes and an extinction position different from that of the related host grains.

Reactivity potential attributed to:
Intensely strained quartz with interlobate boundaries, grain boundary migration, subgrain rotation, and microcrystalline quartz.

Aggregate particle in a sample of concrete from a hydraulic structure (Germany).

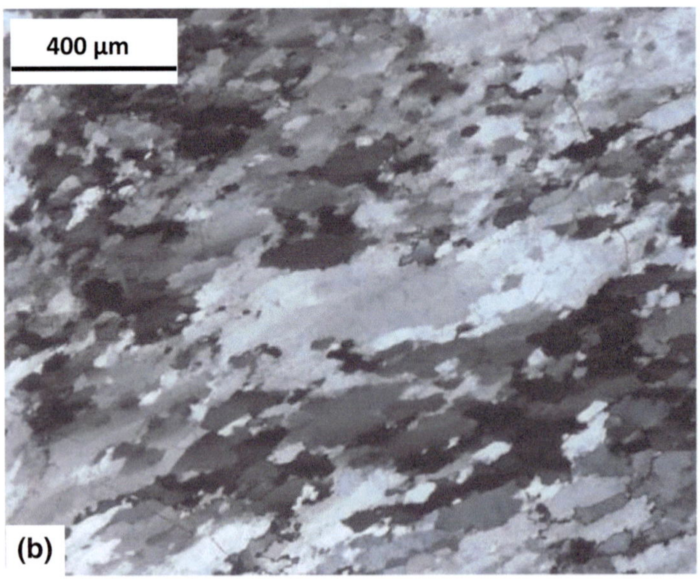

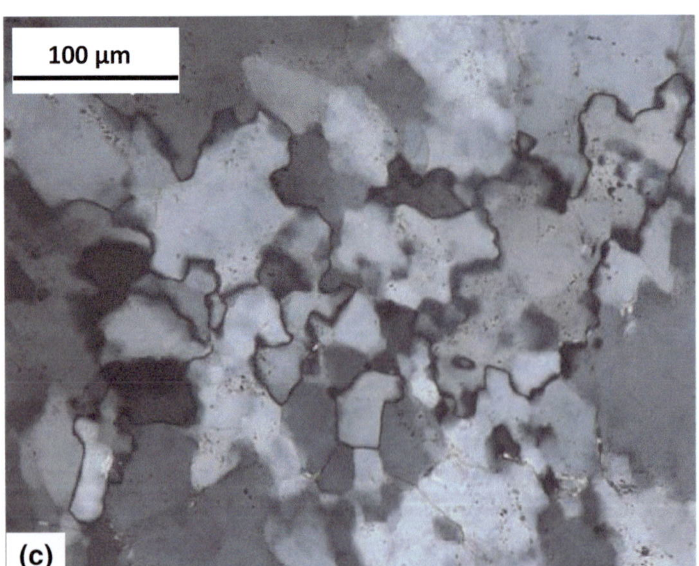

Quartzite (NL)

Deformed rock composed of quartz. It is a quite pure quartzite. The rock is highly deformed and extensively recrystallized with smaller grains along the boundaries of the larger subgrained quartz. Larger grains of quartz show undulatory extinction, deformation lamellae and lobate to serrated grain edges (a, XPL). The stretched crystals define the preferred orientation of the rock (b, XPL). The boundaries of the larger crystals present bulges, grain boundary migration and recrystallized grains (c, XPL).

Reactivity potential attributed to:
Intensely strained quartz with interlobate boundaries and subgraining. Quartz shows bulges, grain boundary migration and recrystallized grains.

Aggregate particle in sample of concrete from a viaduct (The Netherlands).

Quartzite (ZA)

Coarse-grained, impure quartzite composed of quartz, kyanite, muscovite, sericite, chloritoid, zircon, monazite and opaque minerals arranged in a non-directional heterogranular texture. Coarse-grained quartz is the main constituent, with interstitial blasts of kyanite and chloritoid, and white mica flakes along the grain boundaries (a, XPL). Quartz grain boundaries are serrate to interlobate, straight against micas (b, XPL). Larger quartz grains display substantial undulatory extinction and subgraining with formation of microcrystalline quartz, locally recrystallizing to form triple junctions (c, XPL). The sedimentary protolith (e g. detrital sandstone with minor clay content) has been subject to metamorphic conditions leading to formation of kyanite and muscovite/sericite from detrital clay precursors.

Reactivity potential attributed to:
Microcrystalline quartz along the boundaries of the larger clasts. Strained quartz with sutured boundaries. Domains of quartz mosaics due to recrystallization.

Aggregate particle in concrete from a retaining wall (South Africa).

Quartzite (ZA)

Impure quartzite composed of medium to coarse-grained quartz, chlorite and lithic fragments (quartzite). The fine-grained matrix consists of sericite, muscovite and microcrystalline quartz (a, XPL), partly detrital, partly neogenic. Coarse clasts are angular to subrounded and comprise monocrystalline and polycrystalline particles of quartz showing undulatory extinction and deformation lamellae. The lithic fragments (metamorphic rock fragments) show deformed quartz with subgraining (b and c, XPL) and serrate grain contacts, indicating metamorphic provenance. The matrix is composed of sericite from prograde recrystallization of clay minerals under diagenetic to low-grade metamorphism (c, XPL). The low-grade metamorphic features are pointed out by pressure-solution and indentation, with irregular grain boundaries between adjacent quartz grains.

Reactivity potential attributed to:
Microcrystalline quartz in the coarse grains and in the matrix. Subgrains of quartz in the metamorphic rock fragments.

Aggregate particle in concrete from a retaining wall (South Africa).

Quartzite (ZA)

Coarse-grained, impure quartzite mainly composed of quartz, chlorite, opaque minerals, lithic fragments (quartzite) and chloritoid. The rock contains quartz and lithic fragments of a wide size range (a, XPL), with microcrystalline quartz and neogenic chlorite filling interstitial space, locally forming clusters (b, XPL). Some blasts of chloritoid indicate low to medium grade metamorphic conditions. Quartz grain boundaries are generally curved but irregular where bordering on chlorite. Quartz in lithic fragments may have a polygonal fabric (c, XPL).

Reactivity potential attributed to:
Microcrystalline quartz in the intergranular spaces. Quartz crystals with sutured boundaries and bulging.

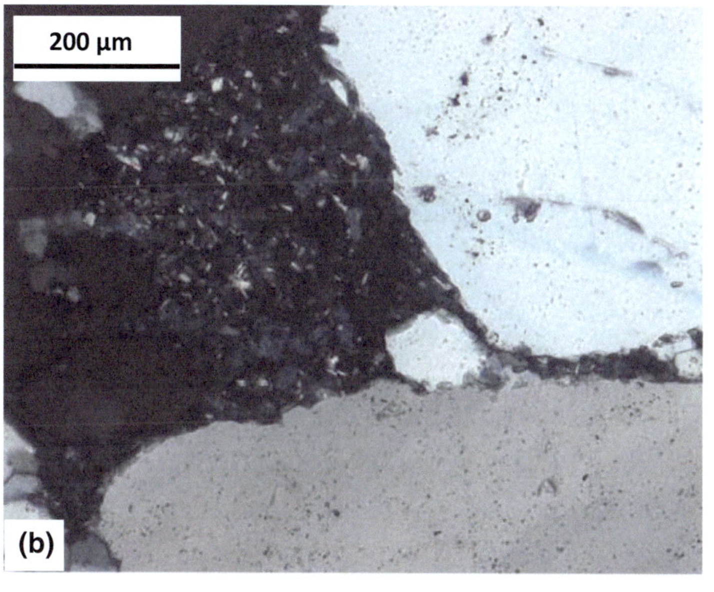

Aggregate particle in concrete from a retaining wall (South Africa).

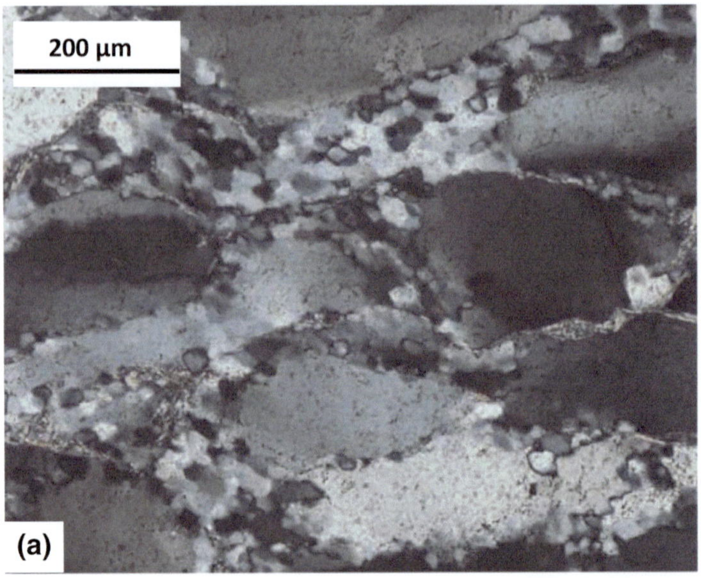

Quartzite (ZA)

Coarse-grained quartzite composed of quartz with trace amounts of sericite, chlorite and opaque minerals. The original rounded clastic particles have been deformed to almond shape (a, XPL), which is accompanied by extensive recrystallization and subgraining forming microcrystalline quartz (b, XPL). Coarse quartz grains show undulatory extinction, and are lined with thin seams of neogenic sericite along their edges (c, XPL). Quartz grain contacts are elongate and curved against sericite mica, serrate to interlobate against other quartz, indicating recrystallization (c, XPL).

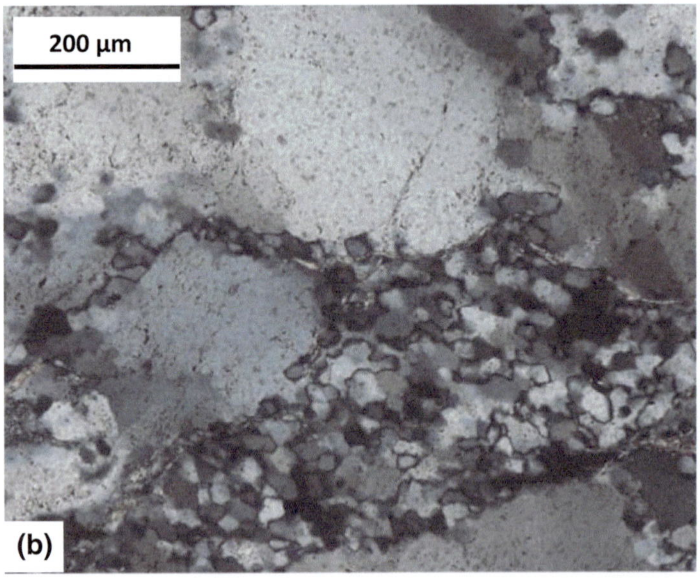

Reactivity potential attributed to:
Subgraining and recrystallization of quartz by dynamic deformation. Abundant microcrystalline and strained quartz.

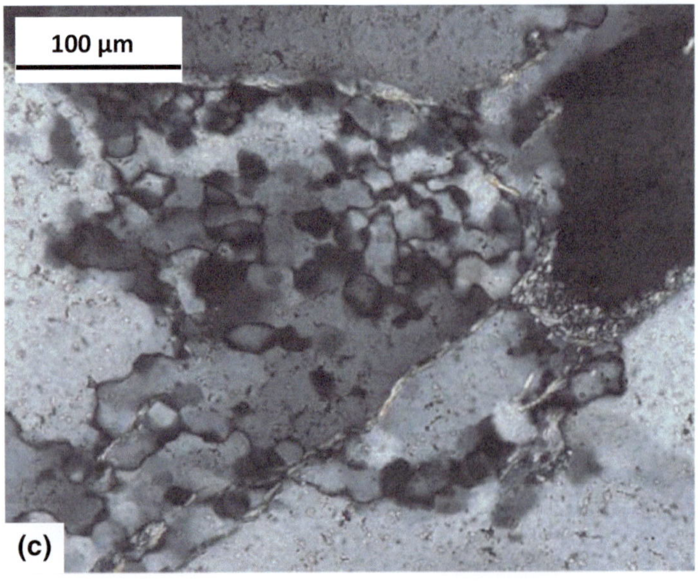

Aggregate sample from a formation classified as potentially reactive (South Africa).

Schist (CH)

Schist composed of quartz, muscovite, biotite, K-feldspar, garnet, chlorite and opaque minerals. Inequant quartz grains and tiny plates of mica display the preferred orientation of the rock (foliation). Porphyroblasts of garnet and K-feldspar are moulded by stretched quartz and phyllosilicate bands (a, XPL). The rock exhibits a well-imbricated grain structure (b, XPL). Strained quartz occurs in agglomerations of crystals, with some microcrystalline quartz, sutured boundaries and recrystallization (c, XPL). Undulatory extinction and deformation lamellae characterise the larger blasts.

Reactivity potential attributed to:
Crystals of stretched quartz with strain lamellae and sutured boundaries. There are alignments of microcrystalline quartz.

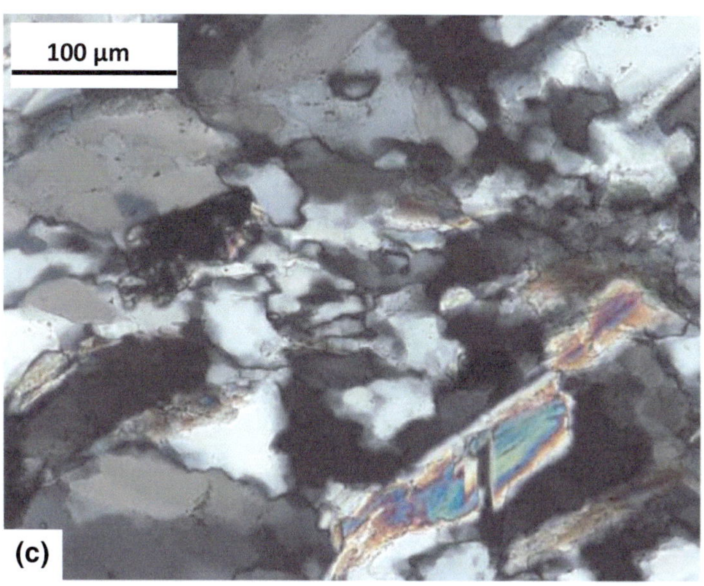

Aggregate particle in a sample of concrete from a dam (Switzerland).

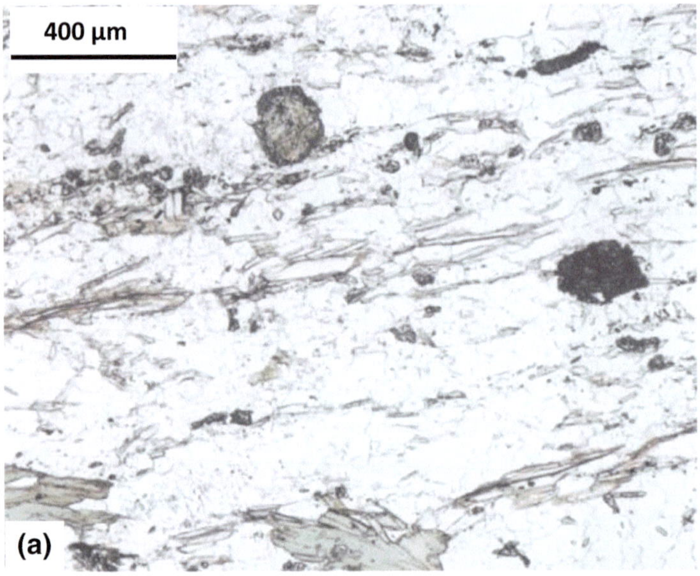

Schist (CH)

Quartz-mica schist composed of quartz, muscovite, biotite, garnet, chlorite and opaque minerals. The rock exhibits well defined penetrative schistosity outlined by the alignment of the plates of chlorite and muscovite (a, PPL). Quartz is the main component and defines a granoblastic texture in which there are inequant crystals of quartz (b, XPL). The crystal boundaries are mainly straight, forming polygonal-shaped mosaics. Stretched crystals of quartz with sutured edges and elongation parallel to the alignment of the micas are observed (c, XPL). The matrix contains microcrystalline quartz.

Reactivity potential attributed to:
Crystals of stretched quartz with sutured boundaries. Microcrystalline quartz in the matrix.

Aggregate particle in a sample of concrete from a dam (Switzerland).

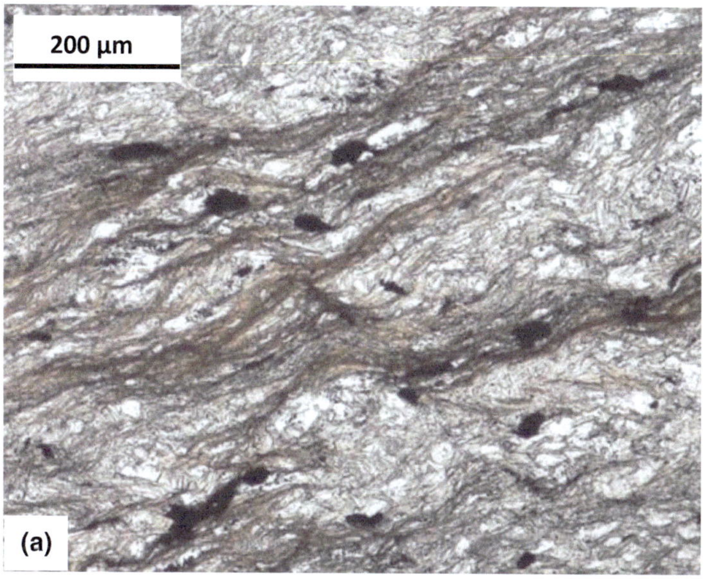

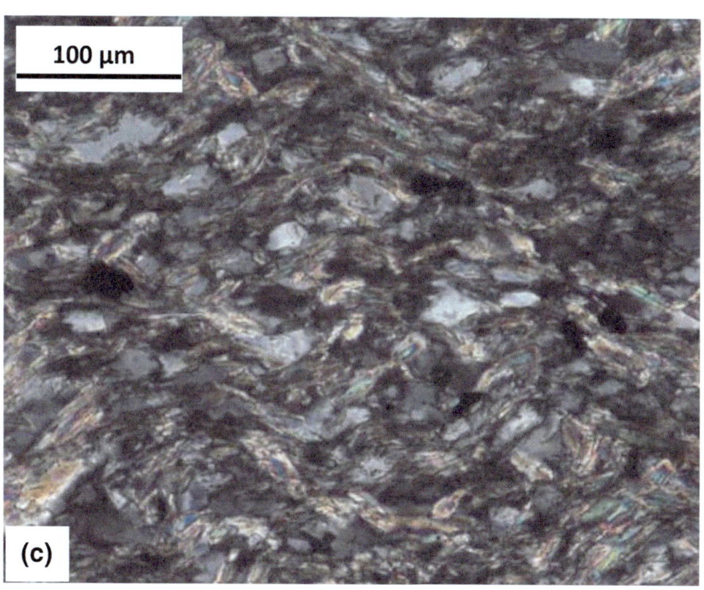

Schist (ES)

Fine-grained schist composed of quartz, muscovite, opaque minerals and iron oxides. The rock displays a penetrative foliation and exhibits a crenulation fabric (microfolds) with an early phyllosilicate foliation folded to produce a new spaced cleavage. The early fabric is still visible (a, PPL). Quartz is segregated in the hinges of the microfolds and phyllosilicates in the limbs, although the quartz occurs dispersed also in the limbs. Quartz crystals follow the alignment of the mica plates, with the longer dimension of the crystals parallel to the mica crystals (b, XPL). Quartz occurs in tiny crystals (microcrystalline quartz) mainly with straight edges. At higher magnification it can be confirmed that quartz crystals are smaller than 50 µm (c, XPL).

Reactivity potential attributed to:
Microcrystalline to cryptocrystalline quartz.

Aggregate particle in a sample of concrete from a dam (Spain).

Schist (ZA)

Schist composed of coarse angular detrital particles of quartz, lithic fragments (quartzite), chloritoid, opaque minerals and muscovite. The quartz displays undulatory extinction and grain contacts are oblong, concave-convex, and locally serrate to interlobate. Lithic fragments appear quartzitic with polygonal internal fabric (a, XPL). Some clasts are overgrown with neogenic quartz, and elongate/flaky clasts follow the foliation (b, XPL). The fine-grained interstitial matrix material consists of microcrystalline quartz, chlorite and sericite formed under metamorphic conditions (c, XPL).

Reactivity potential attributed to:
Microcrystalline quartz in the matrix, recrystallized quartz in rock fragments.

Aggregate particle in concrete from a retaining wall (South Africa).

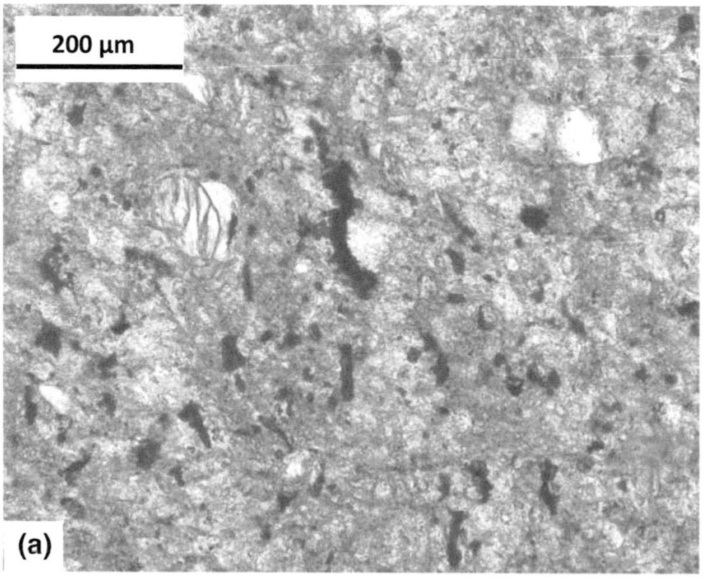

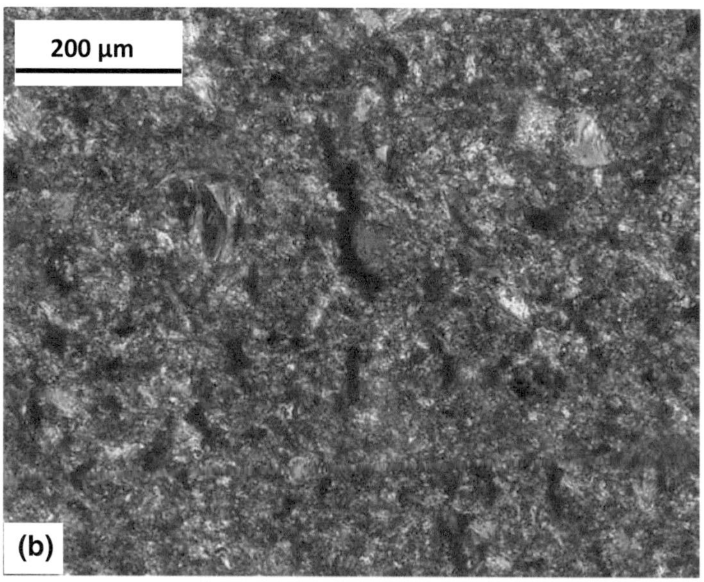

Slate (CN)

Poorly-sorted, medium to fine-grained metasedimentary rock composed of quartz, muscovite and sericite, chlorite, plagioclase, K-feldspar, titanite, and opaque minerals (a, PPL). The rock exhibits an incipient preferred orientation (diagenetic to very-low grade cleavage). Mostly, the rock preserves the original particle shapes, but mineral identity is difficult to distinguish owing to fine grain size (b, XPL). Intercalated intergrowths of muscovite with chlorite imply consolidation under low-grade metamorphic conditions, with neogenic microcrystalline to cryptocrystalline quartz and sericite embedding clastic particles (c, XPL).

Reactivity potential attributed to:
Microcrystalline to cryptocrystalline clastic quartz and neogenic quartz in the matrix.

Sample of rock classified as potentially reactive in laboratory tests (China).

5 Characteristic Examples of Reactive Silica

Isabel Fernandes

Identifying which constituent mineral of a rock is the cause of its reactivity is neither well understood nor evident for many types of rocks. In order to summarise the information already presented in the previous chapters, selected characteristic forms or textures of silica that appear to be susceptible to reaction with the concrete pore solution are illustrated in the following images. These are:

- Pressure-solution
- Strained quartz—undulatory extinction and deformation lamellae
- Subgrains and recrystallization
- Stretched quartz
- Cataclastic quartz
- Pseudotachylite
- Graphic quartz and myrmekites
- Chalcedony
- Volcanic glass

Further information on the petrological background of these silica types can be found in e.g. Graetsch and Grünberg (2012), Katayama et al. (1989), Knauth (1994), Phillips (1974), Passchier and Trouw (2005), Poirier (1985), and some of the definitions in Vernon (2011).

I. Fernandes (✉)
ICT, Institute of Earth Sciences and Department of Geology, Faculty of Sciences, University of Lisbon, Lisbon, Portugal
e-mail: mifernandes@fc.ul.pt

I. Fernandes et al. (eds.), *Petrographic Atlas: Characterisation of Aggregates Regarding Potential Reactivity to Alkalis*, DOI 10.1007/978-94-017-7383-6_5

References

Graetsch, HA, Grünberg, JM (2012): Microstructure of flint and other chert raw materials. Archaeometry (54/1): 18–36.

Katayama, T, StJohn, DA, Futagawa, T (1989): The petrographic comparison of some volcanic rocks from Japan and New Zealand - potential reactivity related to interstitial glass and silica minerals. In: Okada, K, Nishibayashi, S, Kawamura, M (editors): Proceedings of the 8th International Conference on Alkali-aggregate Reaction in Concrete (ICAAR), Kyoto, Japan. CRC Press, Boca Raton/FL: 537–542.

Knauth, LP (1994): Petrogenesis of chert. In: Heaney, PJ, Prewitt, CT, Gibbs, GV (editors): Silica. Physical behavior, geochemistry and materials applications. Reviews in Mineralogy (29): 233–258.

Passchier, CW, Trouw, RAJ (2005): Microtectonics. 2nd edition. Springer Verlag: pp 354.

Phillips, ER (1974): Myrmekite – one hundred years later. Lithos (7/3): 181–194.

Poirier, JP (1985): Creep of crystals: high-temperature deformation processes in metals, ceramics and minerals. Cambridge University Press, Cambridge: pp 260.

Vernon, RH (2011): A practical guide to rock microstructure. Cambridge University Press. ISBN 978-0-521-89133-2. pp 594.

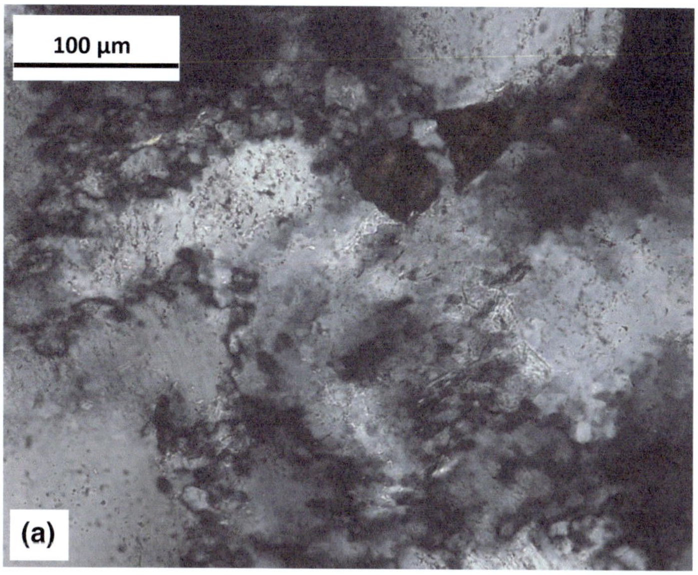

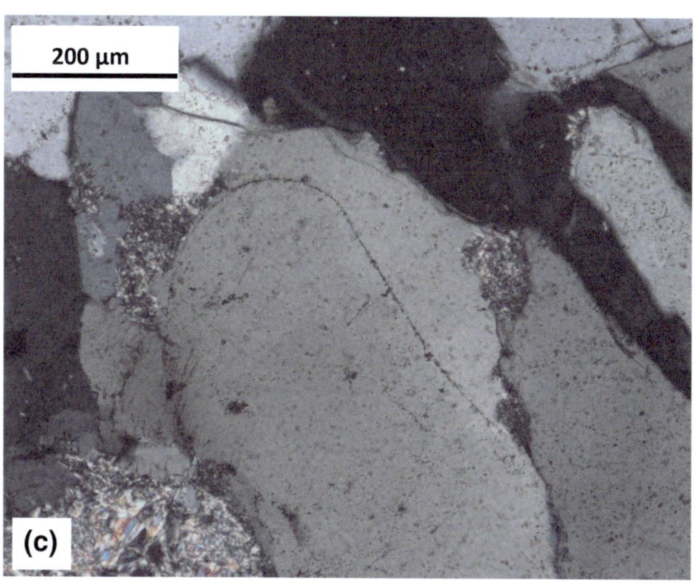

Pressure-solution

Pressure-solution features are formed under conditions typical of deep sedimentary burial in which sediments/rocks are subjected to diagenesis and compaction (in conditions of anisotropic vertical lithostatic pressure) and the start of the metamorphic conditions mainly related to pressure. Silica becomes soluble on the edges perpendicular to the direction of the stress (high-strain areas) and is precipitated in adjacent low-strain areas (a, XPL). These crystals usually develop perpendicular to the main stress direction by deposition of the dissolved material at the boundaries between the grains (b, XPL).

Another process common under diagenesis and compaction conditions is the syntaxial overgrowth quartz in optical continuity with the original detrital grain. The original shape is usually lined by surficial coatings (c, XPL).

a, b (XPL), sandstone, South Africa

c (XPL), sandstone, Canada

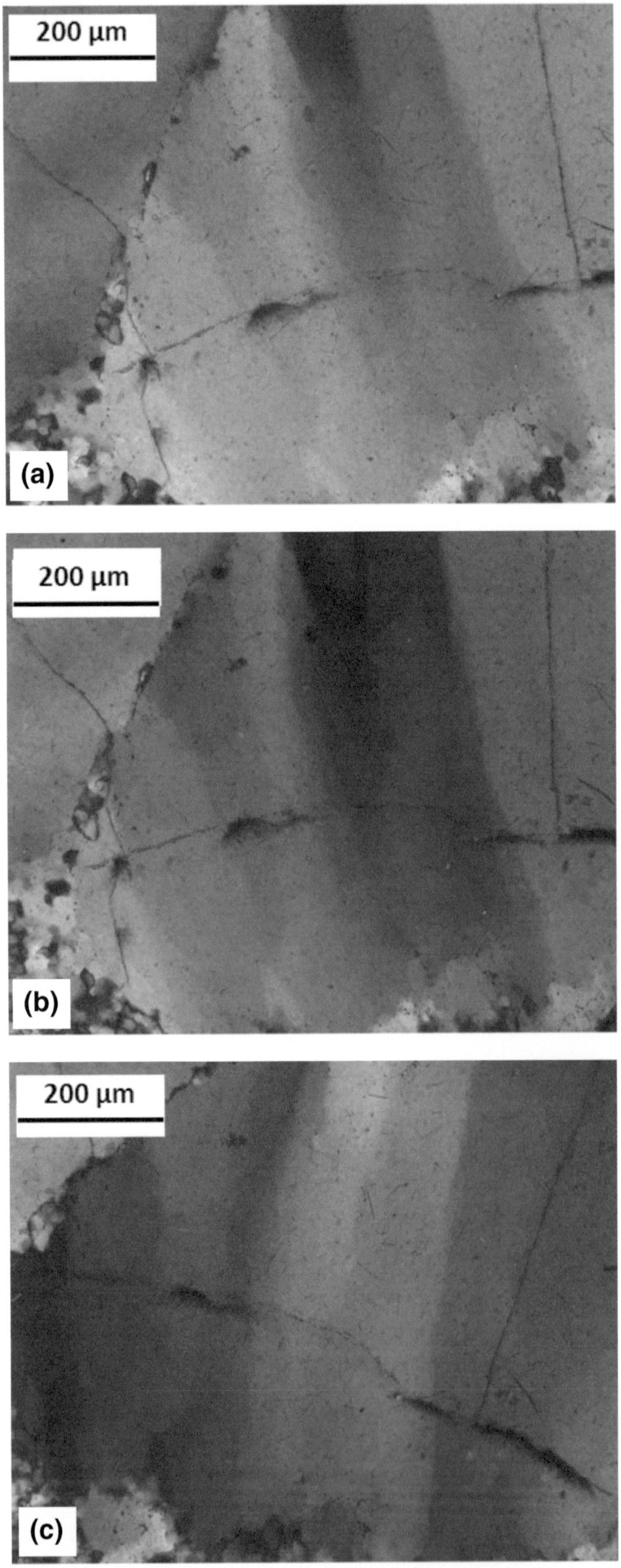

Undulatory extinction (strained quartz)

This crystal of quartz shows sweeping undulatory extinction when rotated (wavy extinction in a single crystal), which indicates the presence of slight bending of the original crystal structure. The boundaries of the crystals are marked by subgrains in alignments. Although there is no simple relationship between the undulatory extinction angle and the degree of reactivity of the aggregates, the presence of strong undulatory extinction is usually associated with subgraining and formation of recrystallized quartz at the grain boundaries.

a, b and c (XPL), mylonite, NO

Deformation lamellae (strained quartz)

Examples of the bands with different extinction timings that constitute the highly deformed crystals. Deformation lamellae can be observed in quartz (a and b, XPL), feldspar (c, XPL) and mica crystals. They are formed by narrow planar, crystallographically oriented zones with refractive index slightly different from that of the adjacent crystal.

a (XPL), porphyry, SE
b and c (XPL), gneiss, FI

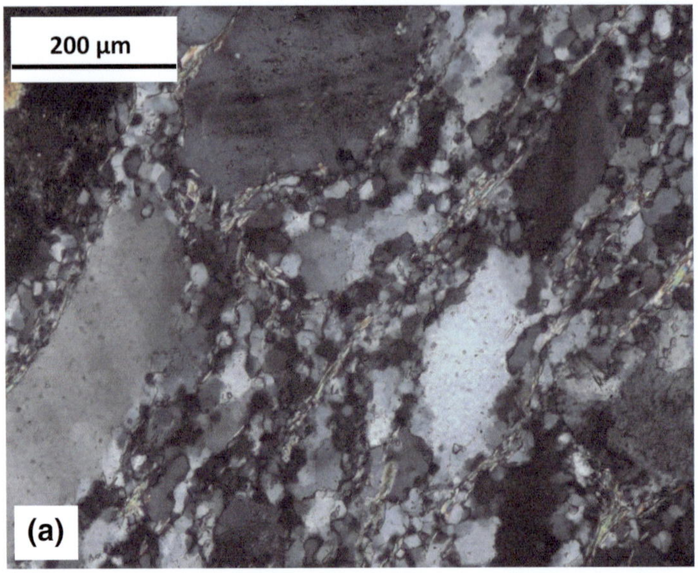

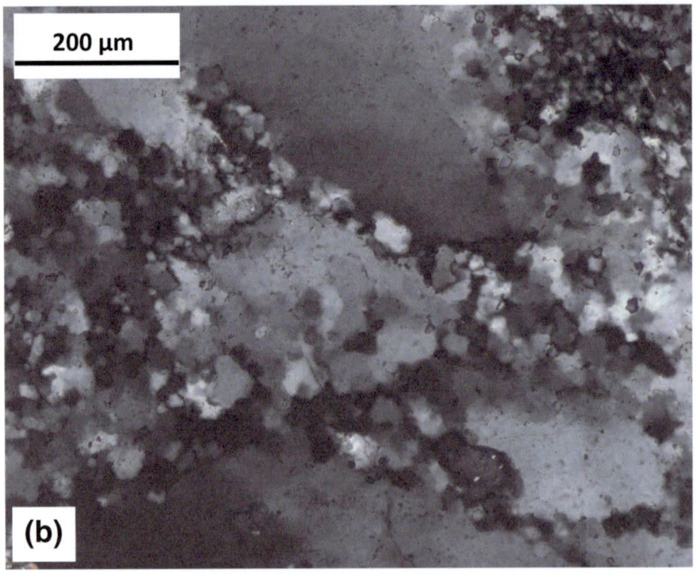

Subgrains and recrystallization

Formation of subgrains (area within an individual and deformed grain that has a slight misorientation of small subgrains) with intense recrystallization (conversion of a mineral grain into an aggregate of smaller new grains of the same mineral owing to deformation) (a, XPL). In these examples, the new grains are <60 µm and some even <10 µm, and they are therefore identified as microcrystalline to cryptocrystalline (incompletely resolved in the optical microscope). The grains are potentially reactive both due to the size and to the sutured boundaries (sutured or bulged, with the grain boundaries interlocking) (b and c, XPL). The occurrence of sutured grain boundaries implies that there is a larger specific surface area potentially exposed to the concrete pore fluids.

a, b and c (XPL), mylonite, NO

Stretched quartz

Elongated aggregations of quartz crystals formed by intense ductile deformation (ribbon structure) (a, XPL). This type of elongated crystal is common in mylonites and other intensely strained rocks such as migmatites and gneisses. The deformation also generates recrystallization with formation of very fine-grained subgrains of quartz (b, XPL). The grain boundaries are irregular as the grains penetrate each other (interlocking texture) (c, XPL).

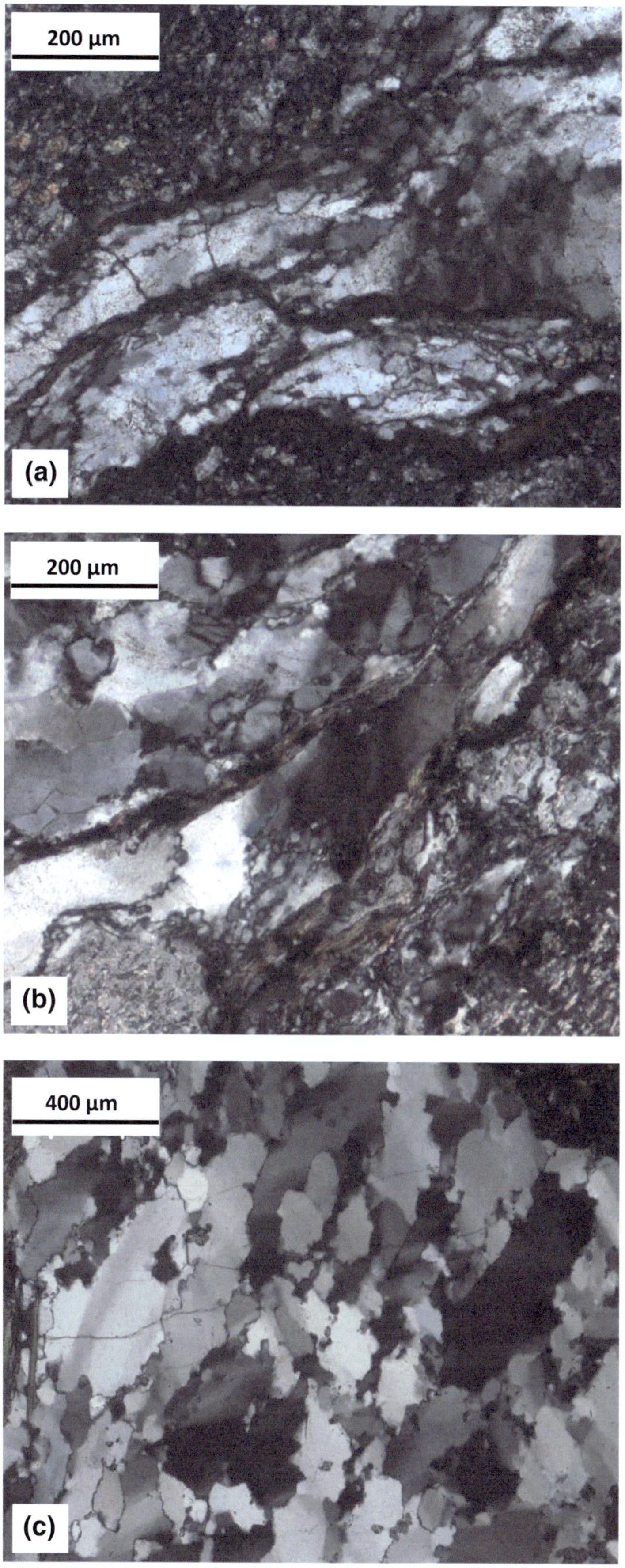

a (XPL), mylonite, NO
b (XPL), mylonite, FI
c (XPL), gneiss, IT

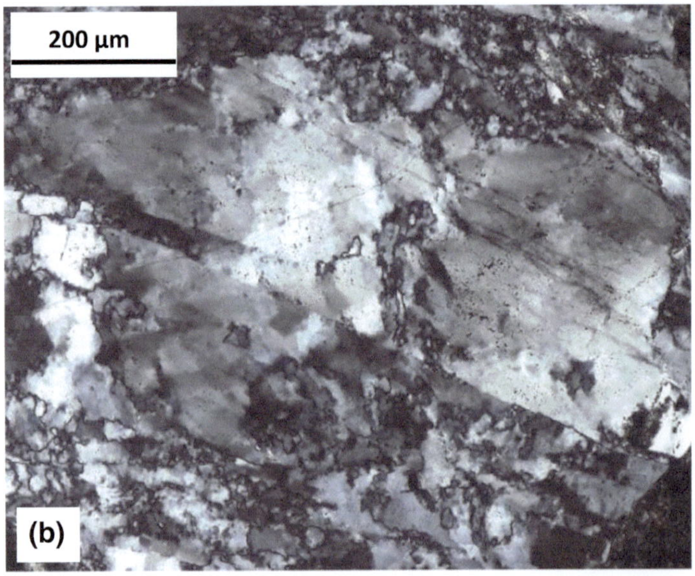

Cataclastic quartz

In contrast to the stretched quartz crystals resulting from ductile deformation, quartz crystals resulting from cataclastic (mechanical) metamorphism are characterised by several cracks and misalignments of crystal features due to brittle behaviour (a, XPL). As a result of the high stress, cracks are usually associated with highly fragmented crystals and also with subgrains (b, XPL). In these types of rock (tectonites or cataclasites) cracks facilitate the access of the concrete pore fluids to the interior of the aggregate particles. The presence of fragmented tiny crystals enhances the potential reactivity. A common feature that is indicative of cataclastic conditions is the effect on the twinning planes of the plagioclase crystals (c, XPL). Planes become undulated and bent, while cracks show frictional sliding along the fracture surfaces of each crystal.

a (XPL), cataclasite, ZA
b (XPL), cataclasite, SE
c (XPL), cataclasite, CN

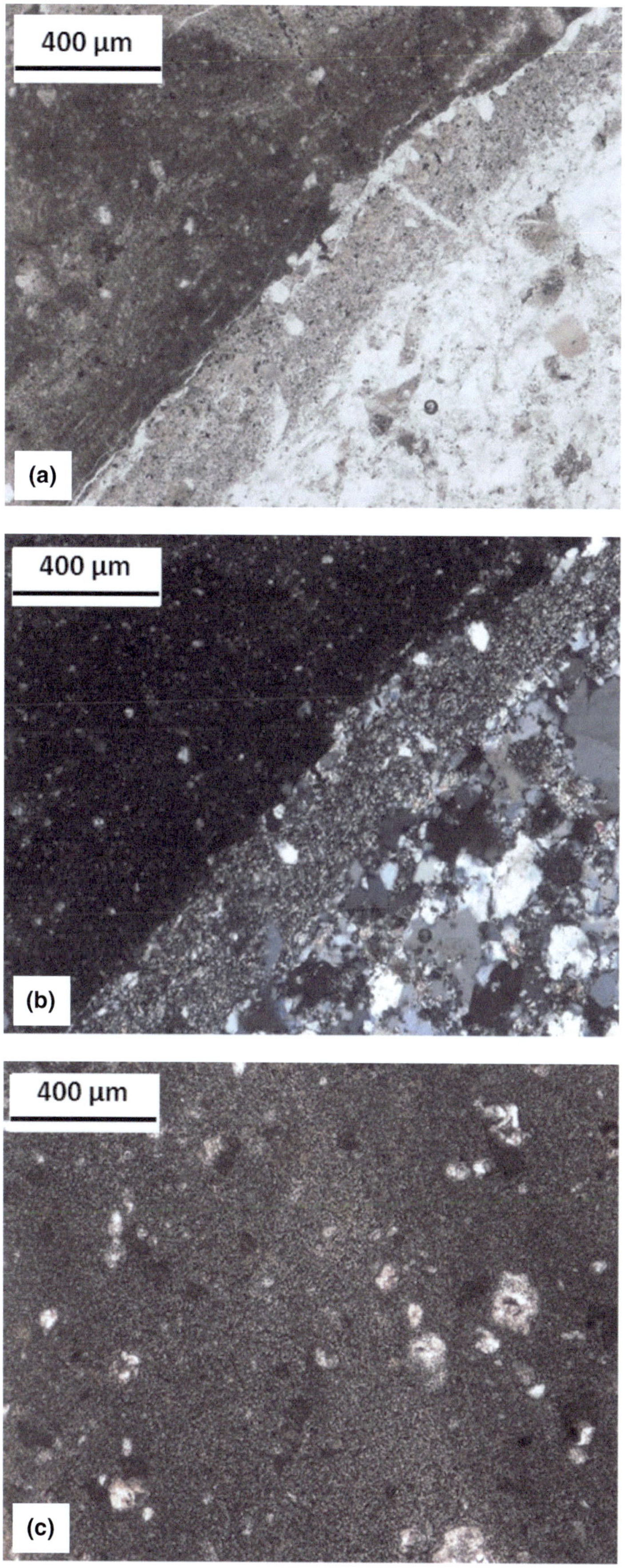

Pseudotachylite

Pseudotachylite is a glassy rock, occurring in veins with sharp contacts with the original rock. It is formed largely by frictional melting along faults in which clasts of the original rock are incorporated. The melted material is vitreous (dark brown areas in the images a, and c, PPL) and therefore isotropic (black in b, XPL).

a (PPL) and b (XPL), pseudotachylite associated with mylonite , SE
c (PPL), pseudotachylite associated with gneiss, AR

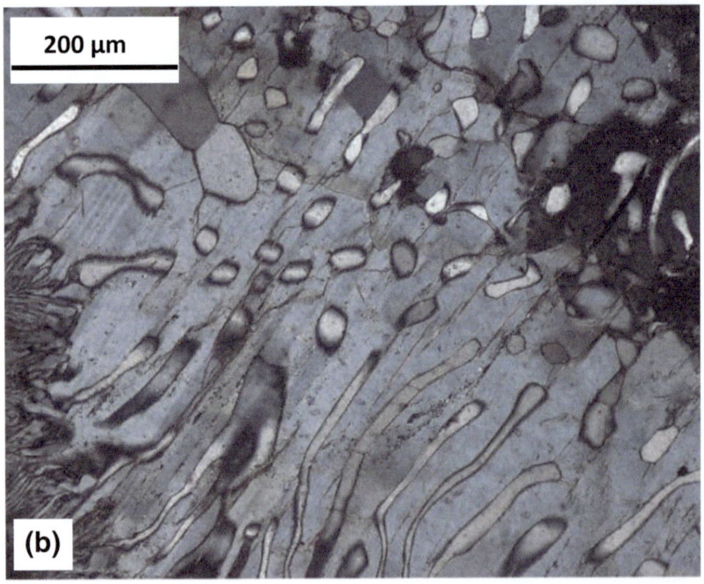

Quartz inclusion and myrmekites

Quartz in microcrystalline inclusions is quite common in igneous rocks, mainly plutonic rocks. The most frequent textures are (vermicular) myrmekites (a and b, XPL), which correspond to symplectic intergrowth, of quartz and sodic plagioclase replacing K-feldspar, in a solid-state reaction. The quartz in the myrmekites is very fine-grained, sometimes even <10 μm (cryptocrystalline). However, its role as a potentially reactive component is still unknown.

Granophyric texture is also common in igneous rocks due to graphic intergrowth of quartz and K-feldspar (c, XPL), each mineral in optical continuity.

a (XPL), felsic porphyry, SE
b (XPL), granite, BR
c (XPL), felsic porphyry, DE

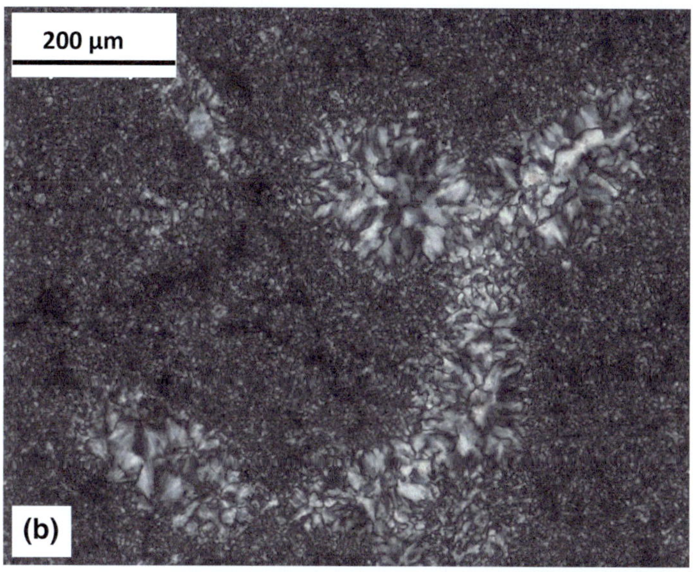

Chalcedony

The feather-like crystals of chalcedony are common in sedimentary rocks such as chert, but can also occur in vesicles of igneous rocks. In sedimentary rocks, chalcedony is often found filling the interior of fossil fragments (a and b, XPL). It is also common in the voids of cherts (c, XPL). As can be observed, these fibrous crystals are extremely small and slim.

a (XPL), chert, NL
b (XPL), chert, FR
c (XPL), chert, DK

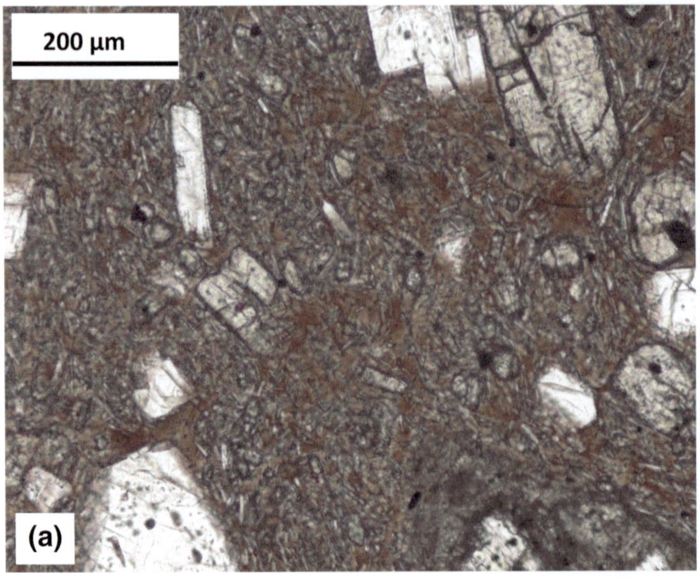

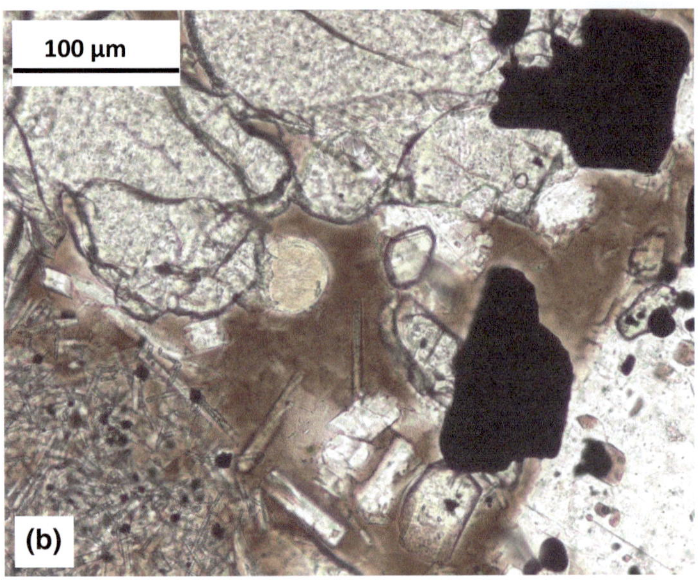

Volcanic Glass

Volcanic glass is formed by rapid cooling of molten lava. It is optically isotropic. Glass is metastable and tends to devitrify (crystallise) with time, becoming potentially more reactive. It can be found in the groundmass of andesite or other volcanic rocks such as basalts and rhyolite. Volcanic glass is not always reactive, depending upon composition and the degree of devitrification. The composition must be assessed by SEM/EDS or by EPMA.

a and b (PPL), andesite, TR
c (PPL), particle manly composed of volcanic glass, concrete prism, IS

Examples of ASR in Selected Rock Types

6

Isabel Fernandes

Features of ASR are presented in the following pages. All the images collected were obtained from the concrete thin sections studied. From the observation of the concrete thin sections, it was concluded that the manifestations of ASR are usually quite discrete in concrete prisms tested under laboratory conditions. In these, cracks are visible only scarcely and the rare spots of gel occur in the cracks that form the interfaces between the cement paste and the aggregate particles. Therefore, some aggregates suspected to be potentially reactive in concrete prisms were not included in the present work, because there was not a positive confirmation of their performance.

By contrast, some of the concretes from structures exhibited exuberant manifestations of ASR: there are cracks crossing aggregate particles, some of which extend into the surrounding cement paste. Some of these cracks are quite wide and most contain alkali-silica gel. The gel is colourless in the cracks and voids formed at the interfaces of the aggregate particles with the matrix, evidencing several desiccation cracks. The gel filling cracks inside the aggregate particles is brownish and very often shows a crystalline structure, sometimes developing a texture similar to rods. Evidence of ASR, namely cracks with gel, are more common in sedimentary rocks such as sandstones and cherts, than in strained rocks e.g. gneisses and mylonites. However, gel was also found in rocks containing "insignificant" contents of microcrystalline quartz in voids and veins.

Owing to the variety of situations in which ASR was observed and the different aspects that can be found even for rocks that are frequently assumed to be innocuous, the present chapter is illustrative and confirms that ASR potential is present in a wide range of rock types. However, it must be emphasised that the evidence of some ASR within concrete is not by itself sufficient to confirm that ASR has caused damage in that concrete (Palmer 1992; Godart et al. 2013). Whilst the presence of alkali-silica gel deposits and evidence of reacting aggregate particles are obviously indicative of some ASR in the concrete in question, reliably linking that reaction to resultant damage is often much more difficult and frequently requires exhaustive examination of a wide range of samples. Direct observation of sites of expansive reaction is essential for confirmation of damaging ASR, together with demonstration of a causal relationship between those sites and the micro-cracking pattern.

References

Palmer, D (Chairman) (1992): The diagnosis of alkali-silica reaction, Report of a Working Party, British Cement Association, Wexham Springs, Slough (now Camberley), UK.

Godart, B, Rooij, M, Wood, JGM (editors) (2013): Guide to diagnosis and appraisal of AAR damage to concrete structures, Part 1: diagnosis. RILEM AAR-6.1, Springer Verlag, Heidelberg-Berlin/DE. RILEM State-of-the-Art Report (12): pp 87.

I. Fernandes (✉)
ICT, Institute of Earth Sciences and Department of Geology, Faculty of Sciences, University of Lisbon, Lisbon, Portugal
e-mail: mifernandes@fc.ul.pt

I. Fernandes et al. (eds.), *Petrographic Atlas: Characterisation of Aggregates Regarding Potential Reactivity to Alkalis*, DOI 10.1007/978-94-017-7383-6_6

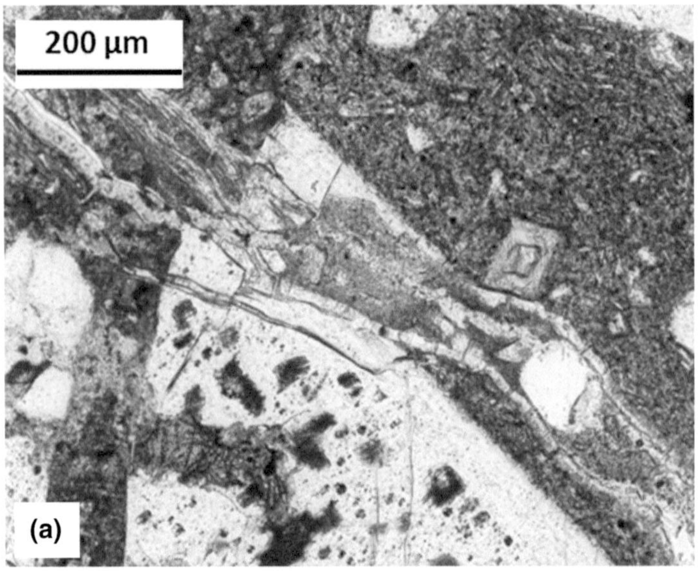

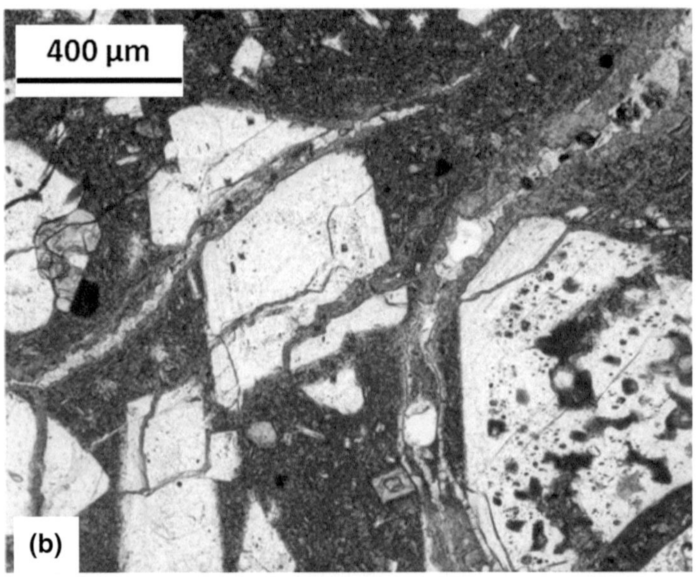

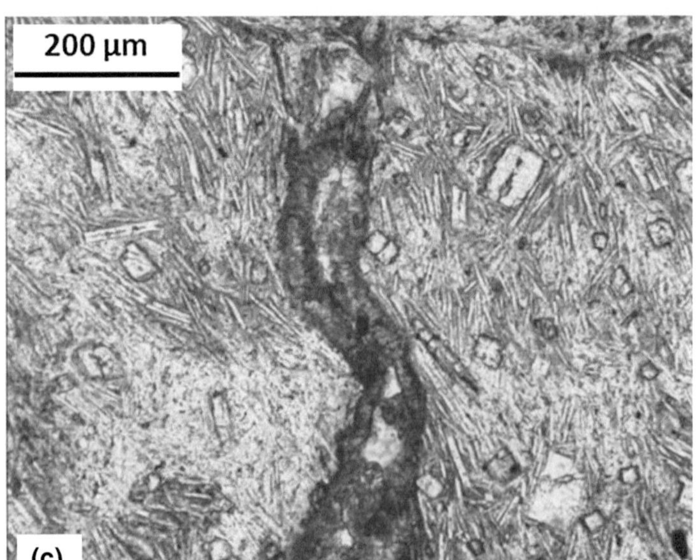

Andesite (JP)

In this thin section of concrete, andesite aggregates contain abundant volcanic glass. The manifestations of ASR are exuberant: wide cracks cross the aggregate particles but also develop at the interfaces with the cement paste (a, PPL). There are cracks in the cement paste which go through the sand particles of aggregate independently of their nature (b, PPL). In a different sample, gel occurs both as ropes but also as rims close to the wall of the cracks. In this case, gel has the texture of radial crystals or rosettes (c,PPL).

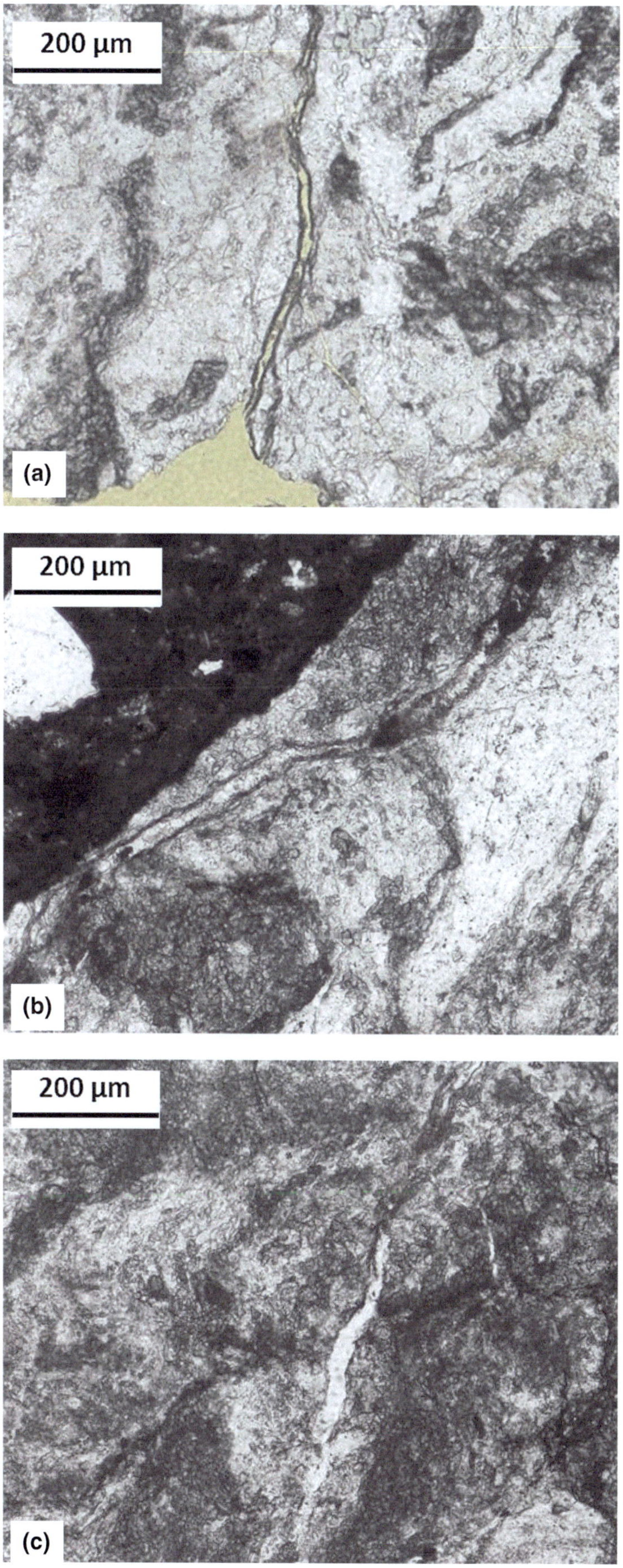

Cataclasite (NO)

The cataclastic calc-silicate rock is composed of tiny quartz clasts and some intensely crushed larger clasts. The cracks cross the rock very often perpendicular to the interface with the cement paste but also sub parallel to the interface (a and b, PPL). There is gel in the cracks located at the interfaces with the cement paste and also some brownish gel in the cracks in the interior of the rock particle (c, PPL).

Note: yellow dye was used in the mounting medium as can be seen in the voids in PPL.

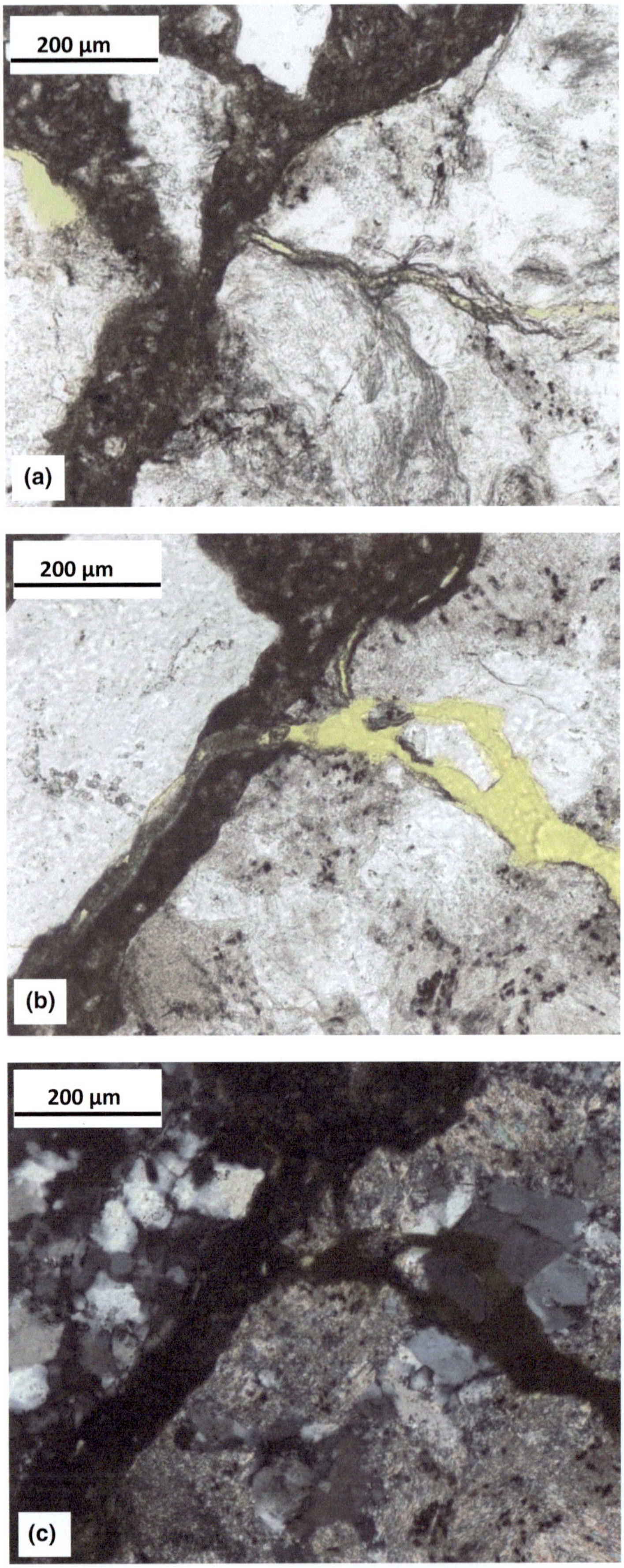

Cataclasite (SE)

The cracks in this sample are mainly perpendicular to the interfaces between the aggregate and the cement paste (a, PPL). They cross the rock and cut the mineral clasts (b, PPL; c, same field of view, XPL). The cracks contain scarce gel in the interior of the aggregate particles, but there is gel filling the cracks when these extend into the cement paste.

Note: yellow dye was used in the mounting medium as can be seen in the cracks and voids in PPL.

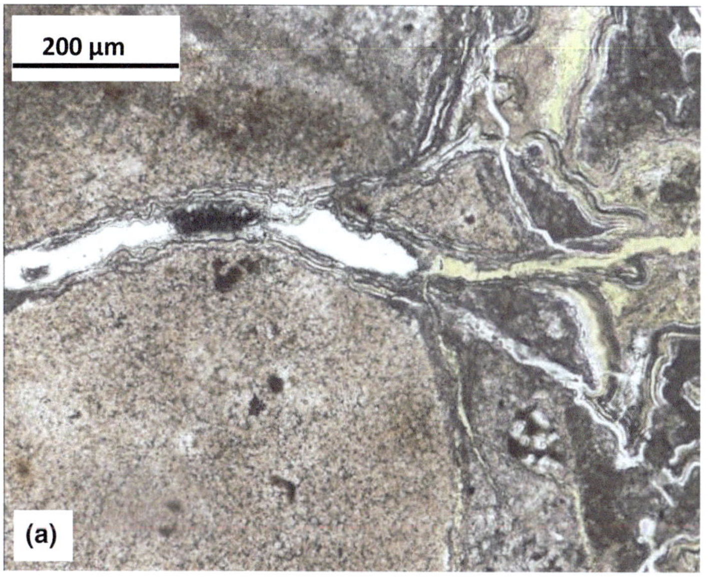

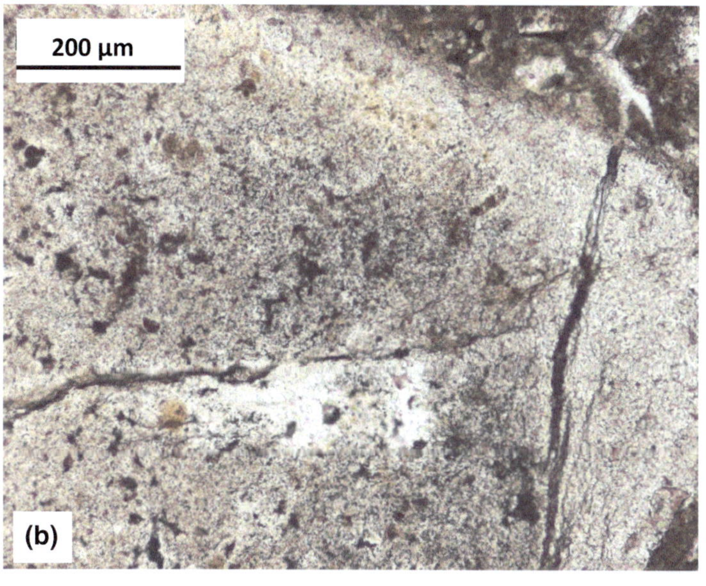

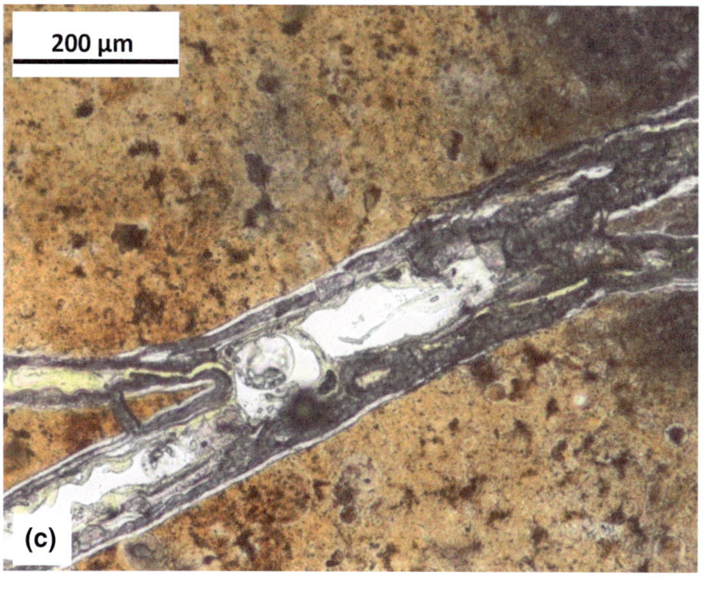

Chert (UK)

In the thin sections containing chert particles, most of the cracks are long and dendritic (with bifurcations) (a, PPL) or thin but crossing the particles (b, PPL). In some places the cracks contain gel lining the walls in layers of different tonality, from brownish to colourless (c, PPL).

Note: yellow dye was used in the mounting medium as can be seen in the cracks in PPL.

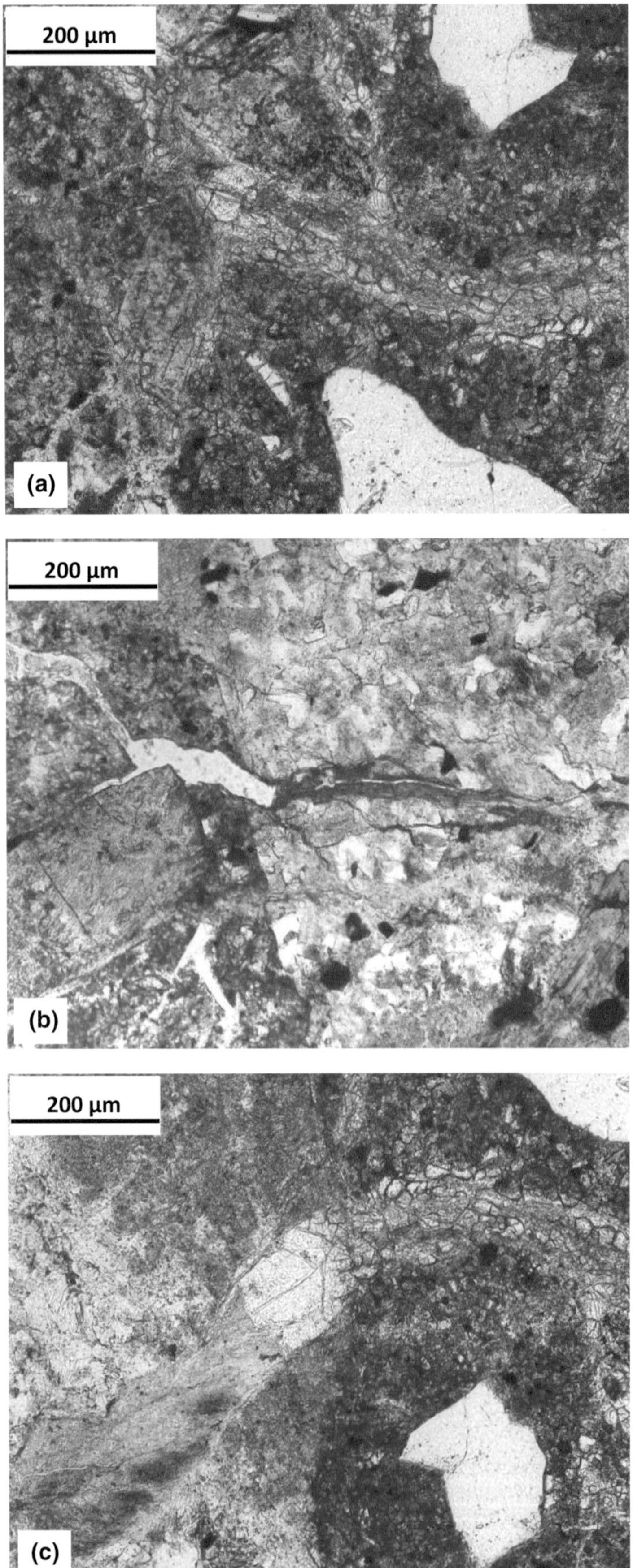

Felsic porphyry (AO)

This felsic plutonic rock is intensely altered: the feldspars are pinkish under PPL. The cracks are numerous and occur both at the interfaces with the cement paste (a, PPL) and crossing the aggregate particle developing perpendicular to the interfaces (b, PPL). These cracks are wide and show abundant gel. The gel is crystalline with a structure similar to ropes (brownish) (c, PPL) or dense, amorphous and colourless, with shrinkage cracks.

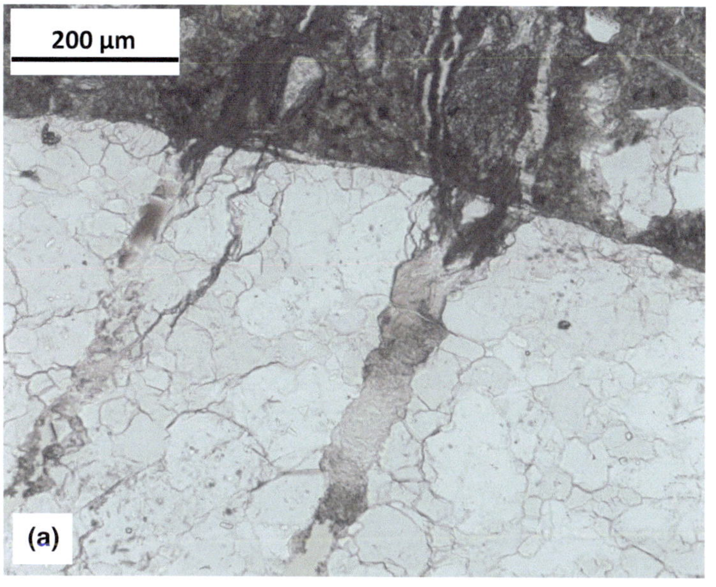

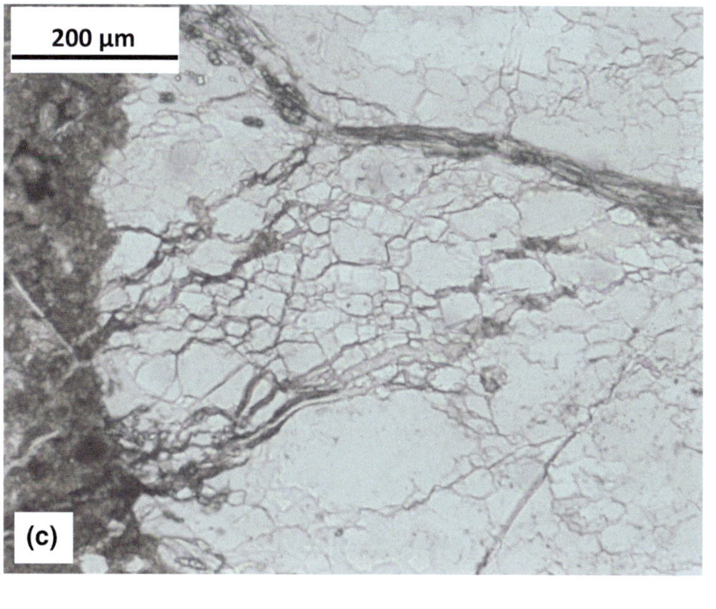

Gneiss (CH)

In this metamorphic rock, the main reactive component is strained quartz . The size of the crystals is usually of the order of (or smaller than) 100µm. The main characteristic is the quite visible contact between every two crystals, which suggests that the intergranular spaces are somehow open and allow the circulation of the concrete fluids into the aggregate particles (a, PPL; b, similar field of view, XPL). The cracks develop perpendicular to the interface with the cement paste (c, PPL). The alkali-silica gel is brownish to colourless, mostly isotropic. There is also some crystalline gel.

Granite (UK)

The samples of concrete with granitic aggregates that have been studied exhibit a particular pattern: cracks show bifurcations (a, PPL; b, same field of view, XPL) and cross the aggregate particles developing perpendicular to the interface with the cement paste (c, XPL). Cracks cut particles of quartz but also of feldspar. However, they seldom contain gel and, when present, it is restricted to the area of the cracks closer to the interface with the cement paste and in the sections crossing the cement paste.

Note: yellow dye was used in the mounting medium as can be seen in the cracks in PPL.

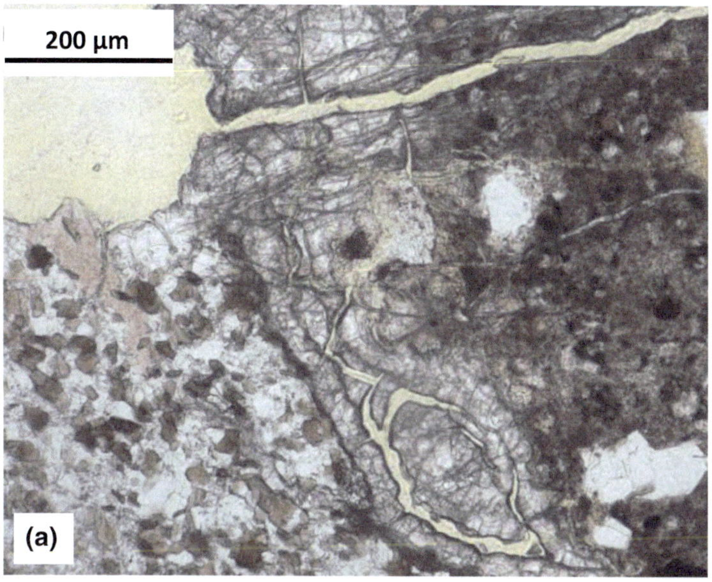

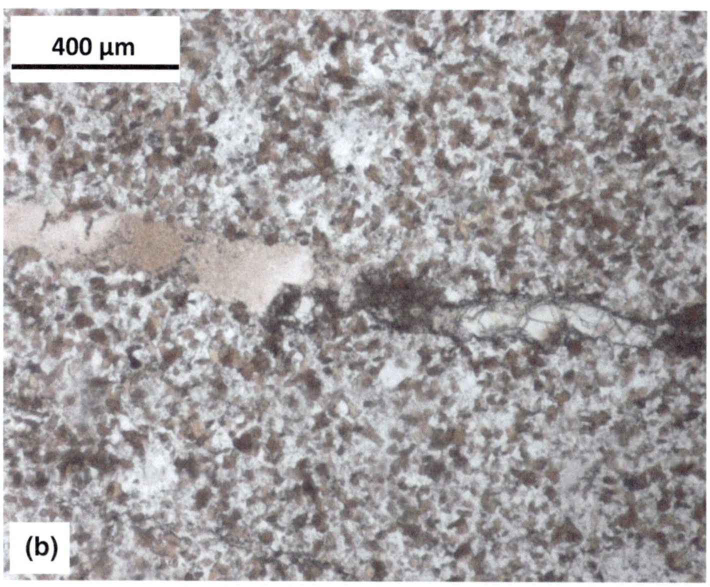

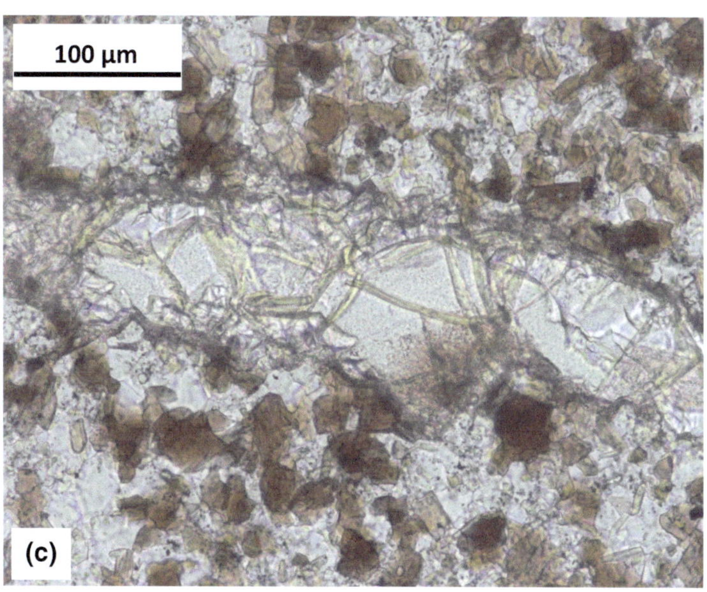

Granofels (FI)

In this very fine-grained metamorphic rock, the cracks show no preferred orientation although most of them develop perpendicular to the interfaces with the cement paste (a, PPL). Gel occurs as a brownish material in the cracks inside the aggregate particles (b, PPL), but dominantly as compact colourless material filling the cracks outside the aggregates and in the interfaces with the cement paste. The cracks have variable width (c, PPL).

Note: yellow dye was used in the mounting medium as can be seen in the voids and cracks in PPL.

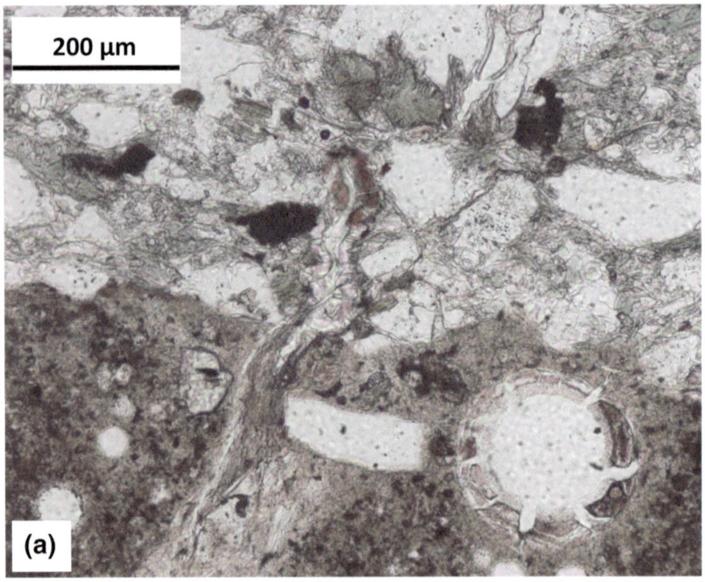

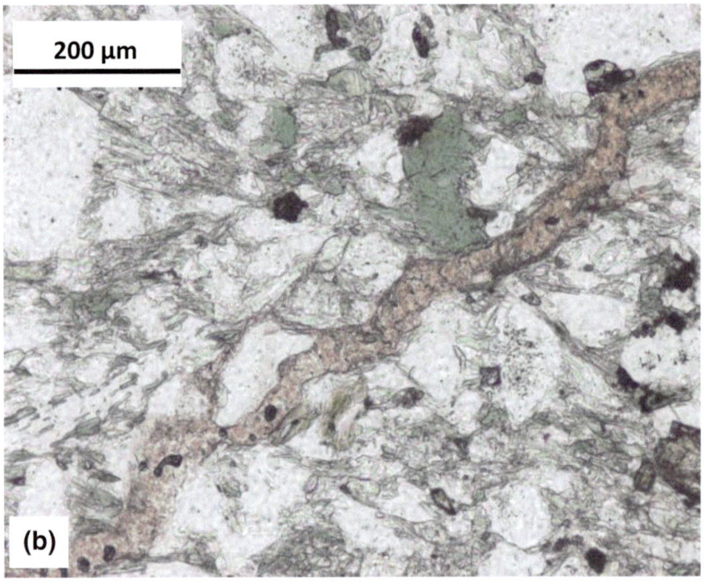

Greywacke (CA)

The cracks in this greywacke occur along the intergranular spaces of the coarser clasts, where chlorite and microcrystalline quartz are dominant (a, PPL). However, there are also cracks cutting the clasts of quartz (b, PPL; c, same field of view, XPL). The gel is brownish, in some places showing crystalline structure.

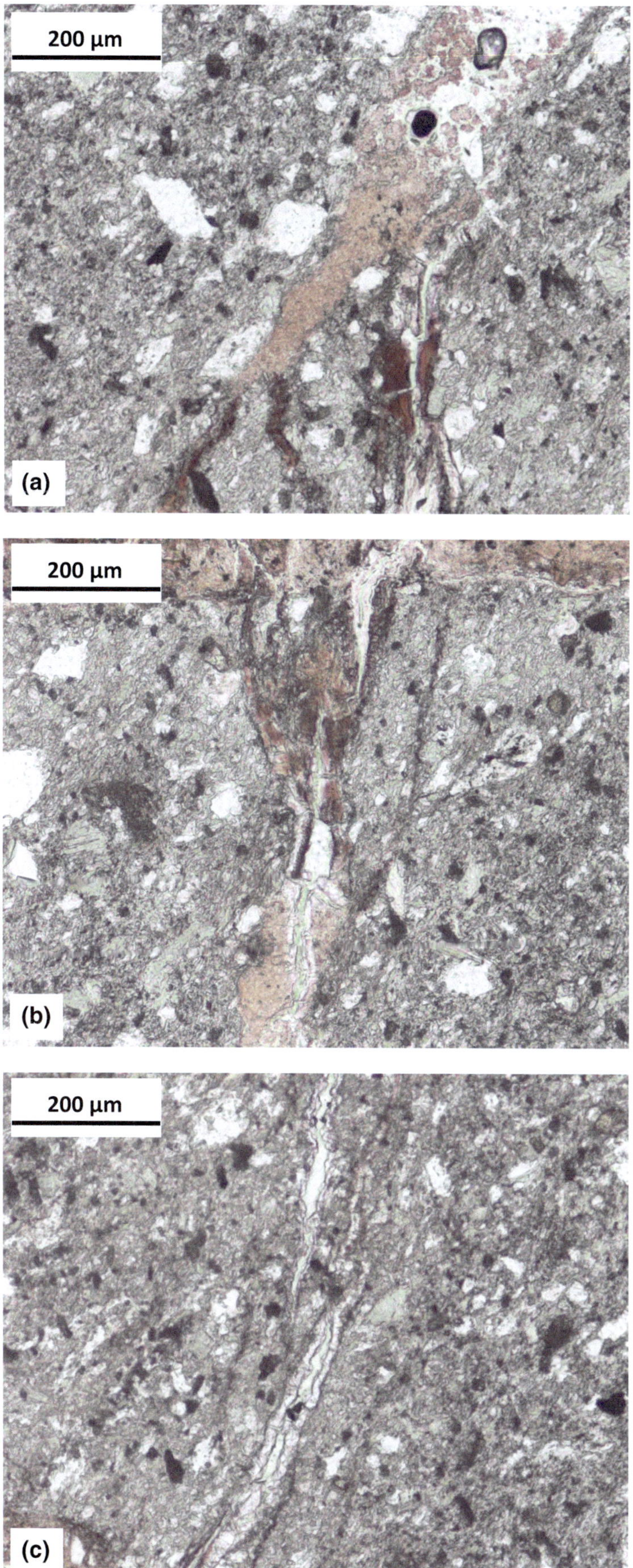

Greywacke (ZA)

In this very fine-grained sedimentary rock, cryptocrystalline quartz and chlorite in the matrix are the most abundant components. The cracks show variable features. Most contain dark brown amorphous gel or brownish crystalline gel (a and b, PPL). The thinner cracks are just lined with colourless crystalline gel with fibrous texture (c, PPL). The cracks extend to the cement paste.

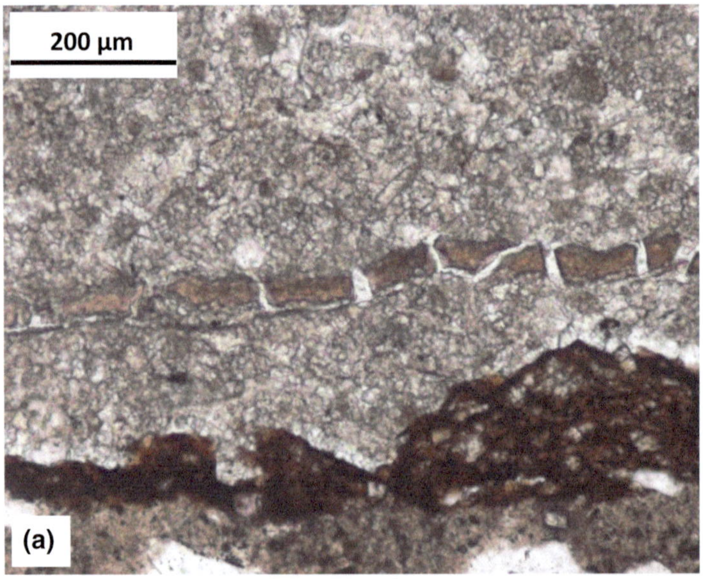

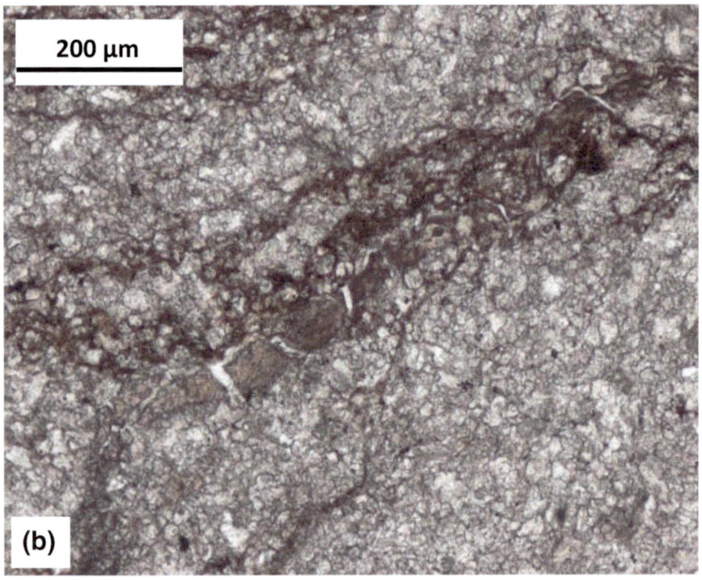

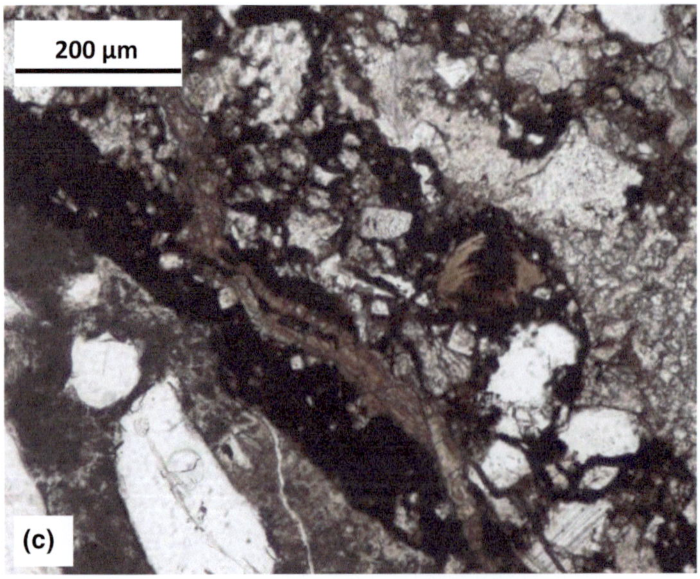

Limestone (CA)

This carbonate rock contains dispersed microcrystalline quartz and also clay minerals. The carbonate minerals form very fine-grained crystals. The cracks observed in most cases develop parallel to the weak lamination of the rock (a, PPL). The cracks contain brownish gel usually with no structure (b, PPL) and occasionally with fibrous character resembling ropes. Most cracks are parallel to the interfaces of the aggregate particles with the cement paste (c, PPL).

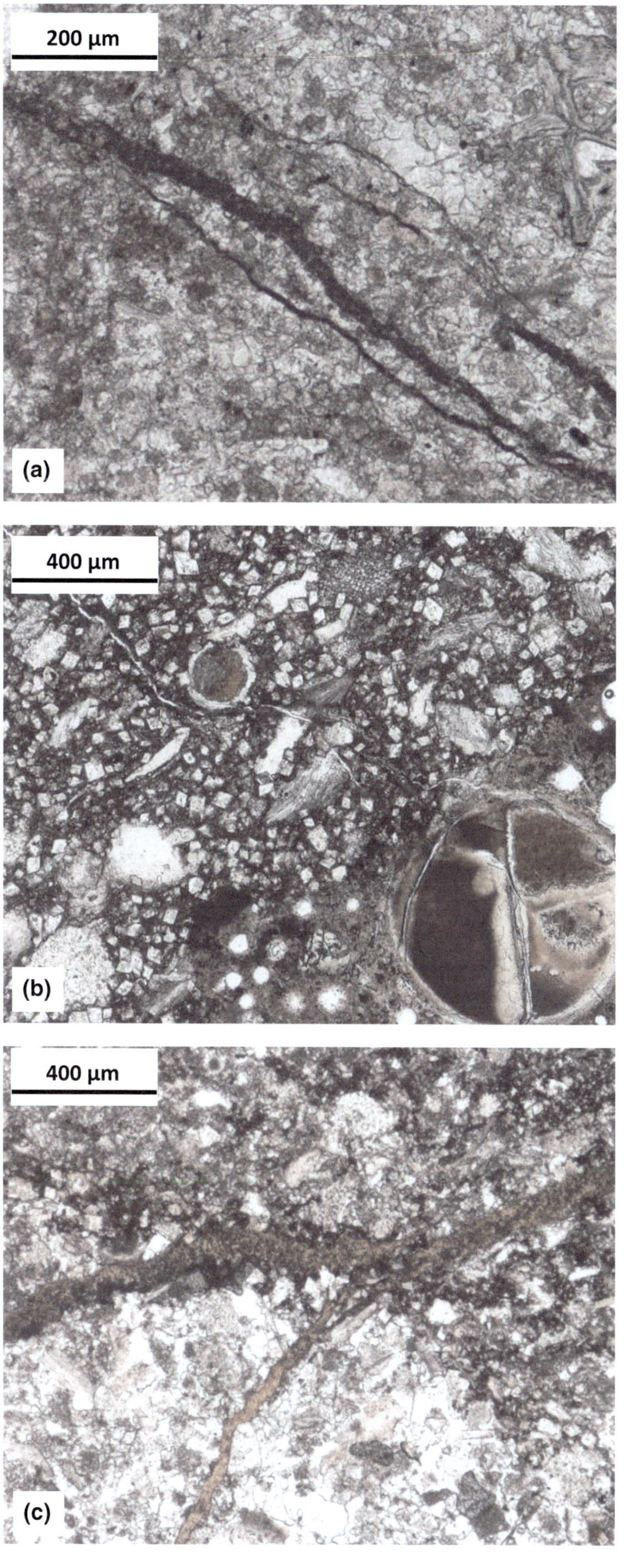

Limestone (CA)

This siliceous dolomitic rock contains fine-grained quartz dispersed in the matrix. There are domains with well-developed crystals of carbonate and other domains in which the matrix is fossiliferous, argillaceous and very fine-grained. Brownish isotropic gel is observed both in the cracks which cross the aggregate particles with random orientation (a, PPL) and in the voids in the cement paste close to the interfaces with the aggregate particles (b, PPL). Cracks and gel are present also in the domain composed of coarse sparitic calcite (c, PPL).

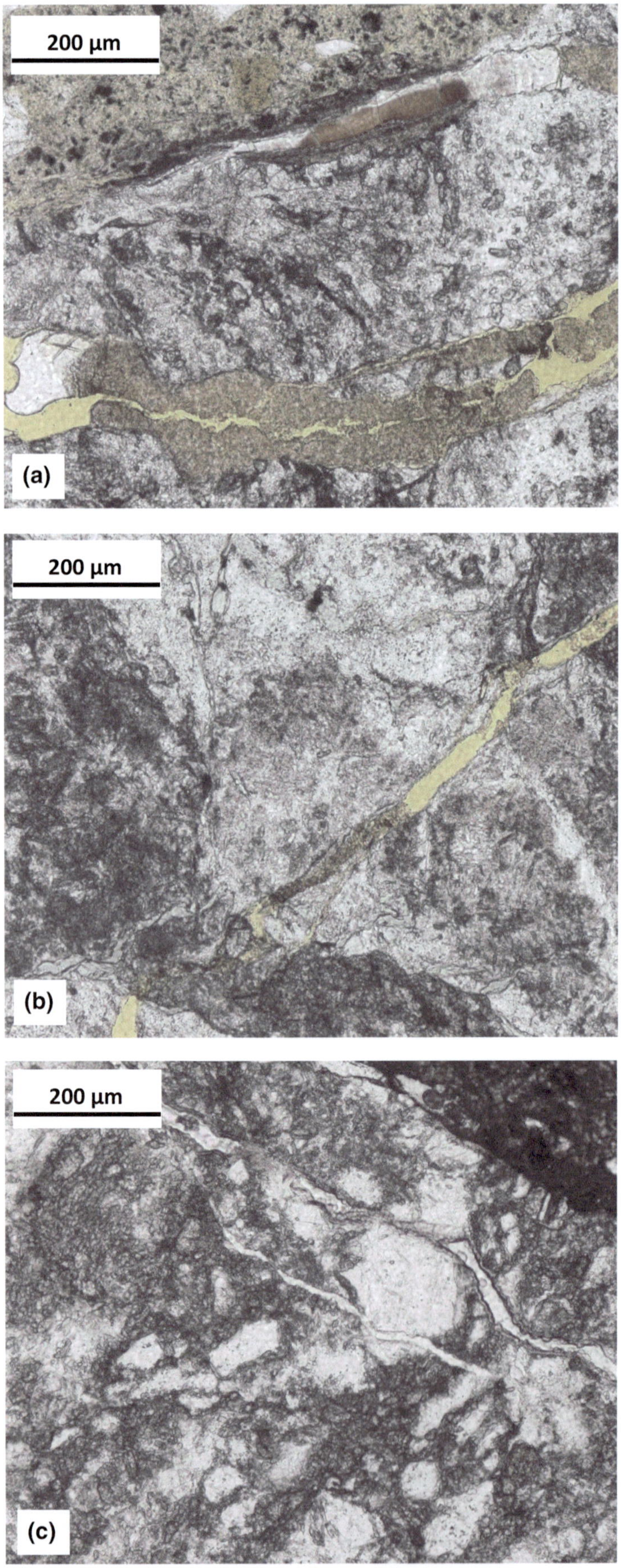

Mylonite (NO)

This calc-silicate mylonite contains highly stretched and microcrystalline quartz. The cracks from ASR show random orientation: some are parallel to the interfaces with the cement paste (a and b, PPL), while others are perpendicular. Usually the wider cracks in this sample contain abundant gel, but in others there is some gel just lining the walls. In some sectors there is no gel (c, PPL). Gel varies from brownish to colourless, locally crystalline.

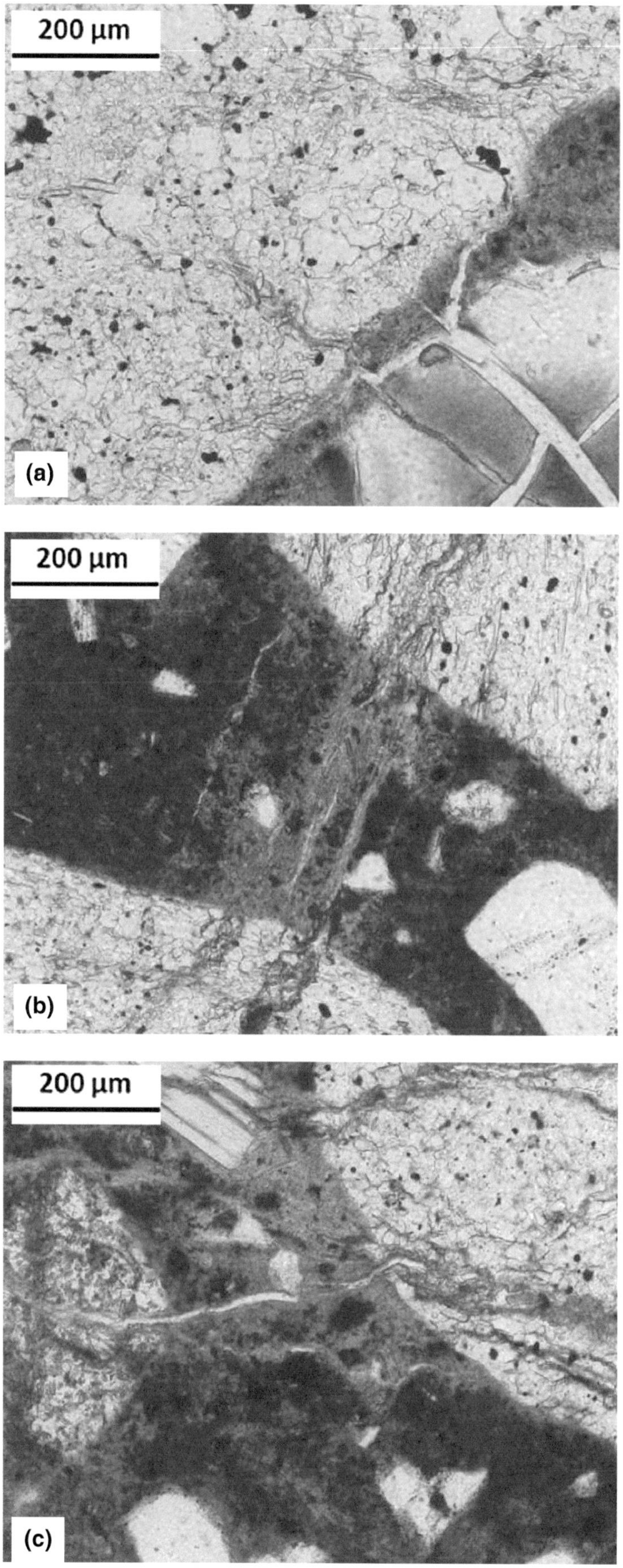

Rhyolite (NO)

The rhyolite shows recrystallization with cryptocrystalline quartz and mica, locally with calcite crystals. The cracks from ASR develop parallel to the alignment of the mica plates and there is abundant gel in voids close to the aggregate particle (a, PPL). However, the cracks also cut megacrysts of quartz or follow the original contacts between crystals. The cracks extend to the cement paste and contain brownish gel both in the interior of the aggregate particles and in the cement paste (b and c, PPL).

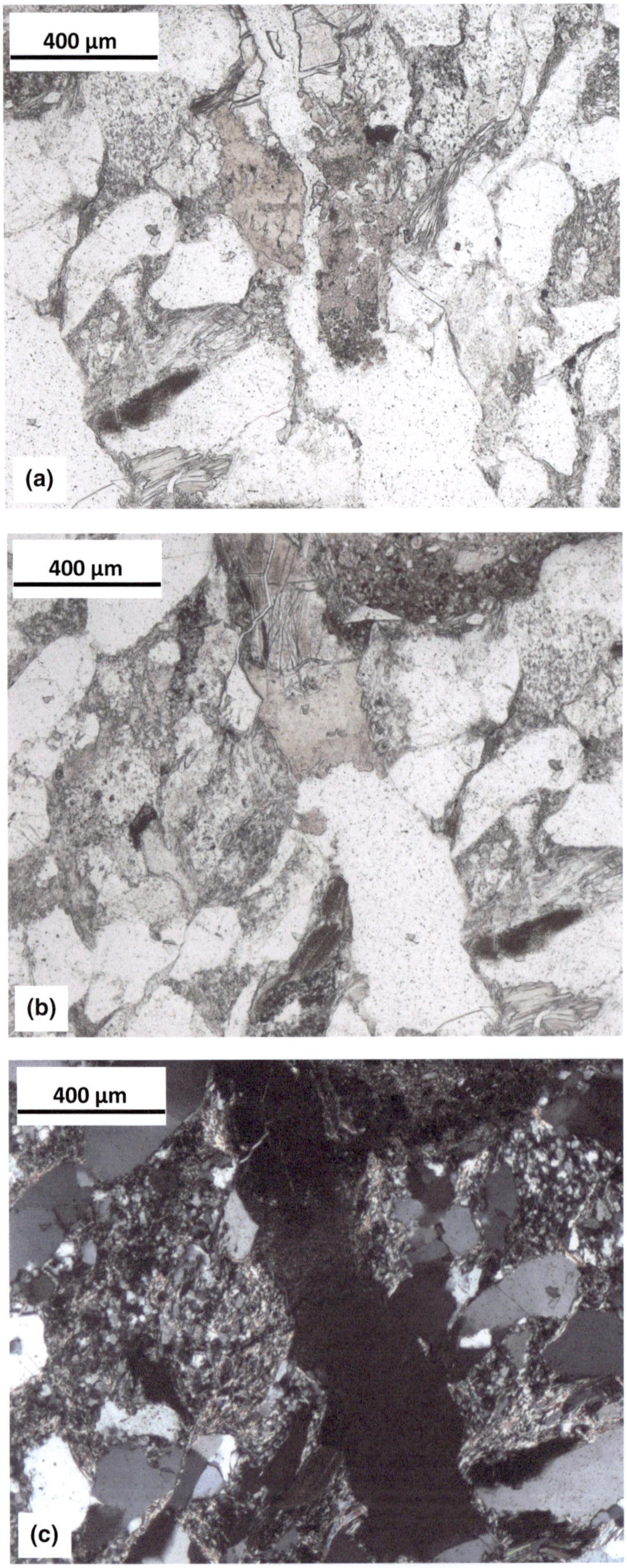

Sandstone (ZA)

This sandstone contains microcrystalline and cryptocrystalline quartz both in the lithic fragments and in the matrix. The ASR cracks are wide and mainly follow the interstitial spaces, developing perpendicular to the particle boundaries. There are also cracks in the interface with the cement paste. Gel is usually brownish to colourless (a and b, PPL). In some sectors of the cracks there is crystalline gel (c, XPL). Gel is present along the cracks both in the interior of the sandstone particle and when crossing the cement paste.

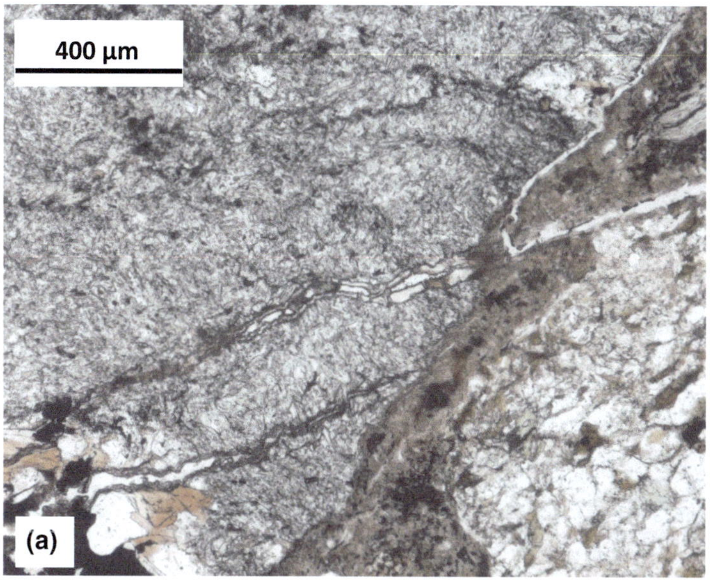

Schist (ES)

Microcrystalline quartz is the main component of this schist. The cracks due to ASR show variable features: in some particles, the cracks align parallel to the foliation defined by the mica plates (a, PPL). Within other particles, the cracks are wider and cross the particles perpendicular to the foliation (b, PPL). In both cases the cracks contain abundant brownish alkali-silica gel (c, PPL).

Bibliography

Adams, AE, MacKenzie, WS (1998): A colour atlas of carbonate sediments and rocks under the microscope. WS Manson Publishing, London: pp 184.

Adams, AE, MacKenzie, WS, Guilford, C (1984): Atlas of sedimentary rocks under the microscope. Longman Scientific & Technical, Harlow/UK: pp 104.

Alaejos, P, Lanza, V (2012): Influence of equivalent reactive quartz content on expansion due to alkali-silica reaction. Cement and Concrete Research (42): 99–104.

Bard, JP (1987): Microtextures of igneous and metamorphic rocks. Revised English edition, translated from French. D. Reidel Publishing Company, Dordrecht/NL: pp 264.

Bérard, J, Roux, R (1986): La viabilité des bétons du Québec: le rôle des granulats. Revue Canadienne de Génie Civil (13): 12–24.

Bérubé, MA, Fournier, B (1986): Les produits de la réaction alcalis-silice dans le béton: étude de cas de la région de Québec. Canadian Mineralogist (24): 271–288.

Bérubé, MA, Fournier, B (1993): Canadian Experience with Testing for Alkali-Aggregate Reactivity in Concrete. Cement and Concrete Composites (15). 27–47.

Best, MG (2003): Igneous and metamorphic petrology. Blackwell Science Ltd. pp 729.

Brandt, I, Schouenborg, B (1990): Guidelines for petrographical micro-analysis of aggregates for concrete. Nordtest—Project no. 826–89. Danish Technological Institute, pp 110.

Cavalcanti, AJCT, Silveira, JFA Degaspare, JC (2000): AAR management at Paulo Afonso IV power plant—Brazil. In: Bérubé, MA, Fournier, B, Durand, B (editors), Proceedings of the 11th International Conference on Alkali-Aggregate Reaction (ICAAR), Québec, Canada: 1263–1272.

Cavalcanti, AJCT, Juliani, M, Iguti, ET, Hahner, I, Zuniga, JEV, (2004): AAR effects at the Paulo Afonso IV power intake. In: Tang, M, Deng, M (editors), Proceedings of the 12th International Conference on Alkali-Aggregate Reaction in Concrete (ICAAR), International Academic Publishers, Beijing World Publishing Corp. Vol. 1: 1035–1042.

Delvigne, JE (1998): Atlas of micromorphology of mineral alteration and weathering. Mineralogical Association of Canada, Special Publication (3): pp 495.

Dolar-Mantuani, L (1969): Alkali-reactive silica rocks in the Canadian Shield, Highway Research Board, Record, 268, 99–117.

Dolar-Mantuani, L (1983): Handbook of concrete aggregates - a petrographic and technological evaluation. Noyes Publications, New York: pp 345.

EN 932-3 (2010): Tests for general properties of aggregates—Part 3: Procedure and terminology for simplified petrographic description. CEN, Brussels, Belgium.

EN 12620:2002+A1 (2008): Aggregates for concrete. CEN, Brussels, Belgium.

Fernandes, I, Noronha, F, Teles, M (2004): Microscopic analysis of alkali-aggregate reaction products in a 50-year-old concrete. Materials Characterization (53/2-4), Special Issue (29): 295–306.

Fernandes, I, Broekmans, MATM, Nixon, P, Sims, I, Ribeiro, MA, Noronha, F, Wigum, B (2013): Alkali-silica reactivity of some common rock types—a global Petrographic Atlas. Quarterly Journal of Engineering Geology and Hydrogeology (46): 215–220.

Fernandes, I, Broekmans, M, Ribeiro, MA, Sims, I (2014): The role of petrographic analysis in the identification of alkali-reactive aggregates. International Multidisciplinary Scientific GeoConferences (SGEM), Vol.II, paper 91 Bulgária: 715–722.

Flörke, OW, Graetsch, H, Martin, B, Röller, K, Wirth, R (1991): Nomenclature of micro- and non-crystalline silica minerals, based on structure and microstructure. Neues Jahrbuch für Mineralogie, Abhandlungen (163): 19–42.

Fournier, B, Bérubé, MA (1991): Évaluation du potentiel de réactivité alcaline des granulats à béton produits dans les Basses-Terres du Saint-Laurent du Québec (Canada). Revue Canadienne de Génie Civil (18/2): 282–296.

Fournier, B, Ideker, JH, Folliard, KJ, Thomas, MDA, Nkinamubanzie, P-C, Chevrierf, R (2009): Effect of environmental conditions on expansion in concrete due to alkali–silica reaction (ASR). Materials Characterization (60): 669–679.

Fournier, B, Nkinamubanzi, P-C, Chevrier, R (2004): Comparative Field and Laboratory Investigations on the Use of Supplementary Cementing Materials (SCM) to Control Expansion Due to Alkali-Silica Reaction (ASR) in Concrete. In: Tang, M, Deng, M (editors), Proceedings of the 12th International Conference on Alkali-Aggregate Reaction in Concrete (ICAAR), International Academic Publishers, Beijing World Publishing Corp. Vol. 1: 528–537.

Freitag SA, Goguel R, Milestone NB (2003): TR 3 Alkali Silica Reaction: Minimising the Risk of Damage to Concrete. Guidance Notes and Recommended Practice. 2nd Edition. Cement and Concrete Association of New Zealand, Wellington, New Zealand.

French, WJ, Howarth, RJ (2000): The petrographic diagnosis of potentially deleterious aggregates. In: Bérubé, MA, Fournier, B, Durand, B (editors), Proceedings of the 11th International Conference on Alkali-Aggregate Reaction (ICAAR), Québec, Canada: 315–324.

Füchtbauer, H, Müller, G (editors) (1970): Teil II. Sedimente und Sedimentgesteine. In: Engelhardt, W, Füchtbauer, H, Müller, G (editors): Sediment-Petrologie. E. Schweizerbart'sche Verlagsbuchhandlung, Stuttgart: pp 726.

I. Fernandes et al. (eds.), *Petrographic Atlas: Characterisation of Aggregates Regarding Potential Reactivity to Alkalis*, DOI 10.1007/978-94-017-7383-6

Gillespie, MR, Styles, MT (1999): BGS Rock Classification Scheme. Volume 1. Classification of Igneous Rocks. British Geological Survey Research Report, 2nd Edition, RR 99-06.

Godart, B, Rooij, M, Wood, JGM (editors) (2013): Guide to diagnosis and appraisal of AAR damage to concrete structures, Part 1: diagnosis. RILEM AAR-6.1, Springer Verlag, Heidelberg-Berlin/DE. RILEM State-of-the-Art Report (12): pp 87.

Goguel, R (1995): Alkali release by volcanic aggregates in concrete. Cement and Concrete Research (25): 841–852.

Grattan-Bellew, PE (1978): Study of expansivity of a suite of quartz-wackes, argillites and quartz-arenites, Proceedings of the 4th International Conference on Alkali-Aggregate Reactions in Concrete (ICAAR), Purdue University, USA: 113–140.

Hoinkes, G, Hauzenberger, CA, Schmid, R (2005): Metamorphic rocks. Classification, nomenclature and formation. In: Selley, RC, Robin, L, Cocks, M, Plimer, IR (editors): Encyclopedia of Geology, Elsevier, Oxford: 386–402.

Jensen, V (1993): Alkali aggregate reaction in southern Norway. Doctor Technicae Thesis, Norwegian University of Science and Technology—NTNU, Trondheim/NO: pp 262.

Johnston, DP, Stokes, D, Fournier, B, Surdahl, R (2004): Kinetic Characteristics of ASTM C 1260 Testing and ASR-Induced Concrete Damage. In: Tang, M, Deng, M (editors), Proceedings of the 12th International Conference on Alkali-Aggregate Reaction in Concrete, International Academic Publishers, Beijing World Publishing Corp. Vol. 1: 338–346.

Joyce, AS (1996): Petrographic aspects of alkali-silica reaction in Eastern Australian concretes. In: Shayan, A (editor), Proceedings of the 10th International Conference on Alkali-Aggregate Reaction (ICAAR), Melbourne, Australia: 767–774.

Katayama, T (1996): Petrography and alkali-reactivity of some volcanic aggregates from Iceland. In: Shayan, A (editor), Proceedings of the 10th International Conference on Alkali-Aggregate Reaction, Melbourne, Australia: 377–384.

Katayama, T (2000): Alkali-aggregate reaction in the vicinity of Izmir, Western Turkey. In: Bérubé, MA, Fournier, B, Durand, B (editors) Proceedings of the 11th International Conference on Alkali-Aggregate Reaction (ICAAR), Québec, Canada: 365–374.

Katayama, T, Futagawa, T (1997): Petrography of pop-out causing minerals and rock aggregates in concrete—Japanese experience. In: Sveinsdóttir, ED (editor), Proceedings of the 6th Euroseminar on Microscopy Applied to Building Materials, Reykjavik, Iceland: 400–409.

Knauth, LP (1994): Petrogenesis of chert. In: Heaney, PJ, Prewitt, CT, Gibbs, GV (editors), Silica. Physical behavior, geochemistry and materials applications. Reviews in Mineralogy (29): 233–258.

Lane, DS (2000): Alkali-silica reactivity in Virginia, USA: occurrences and reactive aggregates. In: Bérubé, MA, Fournier, B, Durand, B (editors), Proceedings of the 11th International Conference on Alkali-Aggregate Reaction (ICAAR), Québec, Canada: 385–394.

Le Bas, MJ, Streckeisen, A (1991): The IUGS systematics of igneous rocks. Journal of the Geological Society of London (148): 825–833.

Lindgård, J, Andiç-Çakır, O, Fernandes, I, Thomas, MDA, Rønning, TF (2012): Alkali-silica reactions (ASR): literature review on parameters influencing laboratory performance testing. Cement and Concrete Research (42): 223–243.

Lindgård, J, Nixon, PJ, Borchers, I, Schouenborg, B, Wigum, BJ, Haugen, M, Åkesson, U (2010): The EU PARTNER Project – European standard tests to prevent alkali reactions in aggregates: Final results and recommendations. Cement and Concrete Research (40/4): 117–128.

Locati, F, Marfil, S, Baldo, E (2010): Effect of ductile deformation of quartz-bearing rocks on the alkali-silica reaction. Engineering Geology (116): 117–128.

Lu, D, Fournier, B, Grattan-Bellew, P (2006): Effect of aggregate particle size on determining alkali-silica reactivity by accelerated tests. Journal of ASTM International (3/9): 1–11.

Lu, D, Fournier, B, Grattan-Bellew, PE, Xu, Z, Tang, M (2008): Development of a universal accelerated test for alkali-silica and alkali-carbonate reactivity of concrete aggregates. Materials and Structures (41): 235–246.

MacKenzie, WS, Adams, AE (1994): A colour atlas of rocks and minerals in thin section. WS Manson Publishing, London: pp 192.

Marfil, S, Batic, O, Maiza, P, Grecco, L, Falcone, D (2010): Comportamiento de rocas basálticas de las Prov. de Corrientes y Entre Ríos frente a la reacción álcali-sílice. VI Congreso Uruguayo de Geología. Uruguay, in CD, pp 6.

Merz, C, Leemann, A (2012): Assessment of the residual expansion potential of concrete from structures damaged by AAR. In: Drimalas, T, Ideker, JH, Fournier, B (editors), Proceedings of the 14th International Conference on Alkali-Aggregate Reactions in Concrete (ICAAR), Austin, USA, pp 10.

Nixon, PJ, Sims, I (2016): RILEM recommendations for the prevention of damage by alkali-aggregate reactions in new concrete structures, State-of-the-Art Report of the RILEM Technical Committee 219-ACS, RILEM State-of-the-Art Reports, Volume 17, Springer (Dordrecht, The Netherlands) RILEM (Paris).

NORMIN 2000 (1999): Hovedprosjektrapport. Alkalireaksjoner I Betong, Bildeatlas. Wigum, BJ (editor) Oslo (in Norwegian).

Paterson, SR (2001): Inclusion trail patterns in porphyroblasts from the Foothills Terrane, California: a record of orogenesis or local strain heterogeneity? Journal of Metamorphic Geology (19): 351–372.

Poole, AB, Sims, I (2016) Concrete petrography - a handbook of investigative techniques, 2nd edition, Taylor & Francis, London, pp 794.

RILEM (2016): RILEM Recommended Test Method: AAR-0 – Outline guide to the use of RILEM methods in the assessment of the alkali-reactivity potential of aggregates. In: Nixon, PJ, Sims, I (editors): RILEM recommendations for the prevention of damage by alkali-aggregate reactions in new concrete structures. Springer Verlag, Heidelberg-Berlin/DE. RILEM State-of-the-art Report (17): 5–34.

RILEM (2016): RILEM Recommended Test Method: AAR-1.1 – Detection of potential alkali-reactivity – Part 1: Petrographic examination method. In: Nixon, PJ, Sims, I (editors): RILEM recommendations for the prevention of damage by alkali-aggregate reactions in new concrete structures. RILEM State-of-the-art Report (17): 35–60.

RILEM (2016): RILEM Recommended Test Method: AAR-7.1 – International specification to minimise damage from alkali reactions in concrete – Part 1: Alkali-Silica Reaction. In: Nixon, PJ, Sims, I (editors): RILEM recommendations for the prevention of damage by alkali-aggregate reactions in new concrete structures. Springer Verlag, Heidelberg-Berlin/DE. RILEM State-of-the-art Report (17): 131–145.

RILEM (2016): RILEM Recommended Test Method: AAR-7.3 – International specification to minimise damage from alkali reactions in concrete – Part 3: Concrete Dams and Other Hydro Structures. In: Nixon, PJ, Sims, I (editors): RILEM recommendations for the prevention of damage by alkali-aggregate reactions in new concrete structures. Springer Verlag, Heidelberg-Berlin/DE. RILEM State-of-the-art Report (17): 155–168.

Shayan, A, Green, WK, Collins, FG (1996): Alkali-silica reaction in Australia. In: Shayan, A (editor), Proceedings of the 10th International Conference on Alkali-Aggregate Reaction (ICAAR), Melbourne, Australia: 85–92.

Shayan, A, Grimstad, J (2006): Deterioration of concrete in a hydroelectric concrete gravity dam and its characterisation. Cement and Concrete Research (36): 371–383.

Shayan, A, Quick, GW (1988): An alkali-reactive basalt from Queensland, Australia. The International Journal of Cement Composites and Lighweight Concrete (10): 209–214.

Shrimer, FH (1996): Evaluation of an alkali-reactive aggregate undetected by petrographic methods. In: Shayan, A (editor), Proceedings of the 10th International Conference on Alkali-Aggregate Reaction (ICAAR), Melbourne, Australia: 734–741.

Sims, I, Nixon, P (2003): RILEM AAR-1 Detection of potential alkali-reactivity of aggregates—petrographic method. TC 191-ARP, Alkali-reactivity and prevention—assessment, specification and diagnosis of alkali-reactivity. Materials and Construction (36): 472–479.

Smaoui, N, Bérubé, MA, Fournier, B, Bissonnette, B, Durand, B (2004): Evaluation of the expansion attained to date by concrete affected by alkali–silica reaction. Part I: Experimental study, Canadian Journal of Civil Engineering (31): 826–845.

St John, DA, Freitag, SA (1996): Fifty years of investigation and control of AAR in New Zealand. In: Shayan, A (editor), Proceedings of the 10th International Conference on Alkali-Aggregate Reaction, Melbourne, Australia: 150–157.

Tiecher, F, Rolim, PH, Hasparyk, NP, Molin, DCCD, Gomes, MEB, Glieze, P (2012): Reactivity study of Brazilian aggregates through silica dissolution analysis. In: Drimalas, T, Ideker, JH, Fournier, B (editors), Proceedings of the 14th International Conference on Alkali-Aggregate Reactions in Concrete (ICAAR), Austin, USA.

Tucker, ME (1991): Sedimentary petrology: an introduction to the origin of sedimentary rocks. 2nd Edition. Blackwell Science, Madden/MA: pp 272.

Vola, G, Berra, M, Rondena, E (2011): Petrographic quantitative analysis of ASR susceptible aggregates for concrete. Proceedings of the 13th Euroseminar on Microscopy Applied to Building Materials (EMABM), Ljubljana, Slovenia.

Wigum, BJ (1995): Examination of microstructural features of Norwegian cataclastic rocks and their use for predicting alkali-reactivity in concrete. Engineering Geology (40): 195–214.

Wigum, BJ, Björnsdóttir, VD, Ólaffson, H, Iversen, K (2007): Alkali-aggregate reaction in Iceland—new test methods. VGK HÖNNUN Consulting Engineers, Report No. VH 2007–036: pp 74.

Yardley, BWD, MacKenzie, WS, Guilford, C (1995): Atlas of metamorphic rocks and their textures. Longman Scientific & Technical, Harlow/UK: pp 126.

MIX
Papier aus verantwortungsvollen Quellen
Paper from responsible sources
FSC® C105338

If you have any concerns about our products,
you can contact us on
ProductSafety@springernature.com

In case Publisher is established outside the EU,
the EU authorized representative is:
Springer Nature Customer Service Center GmbH
Europaplatz 3, 69115 Heidelberg, Germany

Printed by Libri Plureos GmbH
in Hamburg, Germany